Moon Missions

Mankind's first voyages to another world

by **William F. Mellberg**
Plymouth Press, Ltd

© 1997 William F. Mellberg
ISBN 1-882663-12-8

Plymouth Press, Ltd.
42500 Five Mile Road
Plymouth, MI 48170-2544

Edited by Jan Jones

Editorial assistance by Ashley Andersen

All Plymouth Press books are available at a significant discount when purchased in bulk quantities for educational or promotional purposes. Special promotional editions can be formulated to fulfill specific needs. Call our marketing manager at (800) 350-1007 for further information.

Printed in South Korea

Front cover: A precision landing allowed Apollo 12 astronauts to walk to the Surveyor 3 spacecraft which had landed on the Ocean of Storms 31 months earlier. Here Pete Conrad reaches toward Surveyor's television camera, which he and Al Bean later removed and brought back to Earth. The Lunar Module *Intrepid* can be seen on the rim of Surveyor Crater at the upper right of the photo. *(NASA)*

Title page: The Moon provided a backdrop for this dramatic nighttime view of the Saturn V rocket which sent Apollo 17 to the Valley of Taurus-Littrow. *(NASA)*

Back cover: Clockwise from upper left: Soviet Vostok launcher (also on page 28); the Lunar Landing Training Vehicle (also on page 52); Apollo 11 blasting off for the first manned lunar landing (also on page 34); the Lunar Roving Vehicle used by the astronauts for lunar exploration (also on page 118); the shattered Apollo 13 Service Module (also on page 109); and Buzz Aldrin descending from the Apollo 11 Lunar Module to become the second man on the Moon (also on page 93). Central picture of the Moon is a composite of images, copyright UC Regents Lick Observatory.

Dedication

*To the hundreds of thousands of scientists and engineers around the world
who made the Moon missions possible. And to my father, Frank W. Mellberg,
who was one of them.*

Contents

DR. HARRISON H. SCHMITT, APOLLO 17 ASTRONAUT

Foreword *by Harrison H. Schmitt*

I would like to tell you about a place I have seen in the Solar System called the Valley of Taurus-Littrow near the coast of the great frozen sea of Serenitatis. Surrounded by towering walls and filled with countless wonders, this beautiful place was my "home" for three remarkable days in December, 1972. The Valley has been unchanged by being named by men, while change has come to the men who named it. And the Valley has been less altered by being explored than they who have been the explorers.

With a lunar map and a modest telescope, you can see the Valley of Taurus-Littrow in the upper right-hand quadrant of the Full Moon. It stands out as a small, dark indentation in the rim of Mare Serenitatis near the crater, Littrow. As a member of the Apollo 17 crew, I enjoyed the privilege of visiting this breathtaking valley and investigating its environment on one of history's greatest voyages of discovery.

In *Moon Missions*, Bill Mellberg describes man's incredible journeys to the Moon and provides a concise look back at the early history of lunar exploration. But I am pleased that Mellberg also looks to the future. The Moon is not a distant celestial neighbor but rather, with Earth, forms one environmental system that could greatly affect — and improve — our everyday lives. This remains one of the largely untold stories of the Apollo Program. So let me begin by sharing a few thoughts about tomorrow.

It is safe to say that in the absence of some clear action by our civilization, fossil fuels will continue to be the world's dominant source of energy. The total consequences of this truth are uncertain, but it is safe to say that the environmental, political, and military changes arising from this dependence will certainly be adverse, if not catastrophic. With warnings of potential political and environmental danger already spreading across the globe, the prudent course would be to accelerate the search for an economically and environmentally viable alternative to fossil fuels.

That search is once again pointing our sights toward the Moon, since a commercially viable alternative to fossil fuels for the generation of electrical power appears to exist there in the form of solar helium-3. This is a nonradioactive, light isotope of helium, blown out from the Sun's interior, and accumulated in the lunar soil over billions of years. Helium-3 can be used as a highly efficient fuel for fusion power, a fuel with few or no adverse environmental consequences.

The fusion of helium-3 with deuterium, a heavy isotope of hydrogen that is available on Earth, produces large amounts of energy with little radioactive waste. Furthermore, the energy of helium-3 fusion can be converted to electricity at twice the efficiency of existing thermal power plants and of other types of fusion plants now proposed. The fusion of helium-3 with itself, a possible but more difficult process to achieve, produces absolutely no radioactive waste.

Helium-3 was discovered in the lunar soil samples brought back to Earth from the Apollo missions. Although present in concentrations of only 20 to 30 parts per billion, it is far more accessible on the Moon than on our own planet. An almost limitless supply is believed to exist in the upper several meters of pulverized rock on the lunar surface.

Relatively small amounts of helium-3, mined and processed on the Moon, then transported to Earth for use in fusion power plants, could produce enormous amounts of energy. If the electricity now being used in the United States each year were generated by helium-3 fusion, the required fuel could be carried in the cargo bay of a single Space Shuttle flight! The conclusions of financial analyses by a number of scientists support the viability of a helium-3 mining business. Helium-3 fusion also offers the potential of less expensive electrical power for consumers. Another benefit of replacing oil-burning power plants with fusion plants would be global reductions in the emission of pollutants.

This is not a fantastic dream. It is a viable commercial option to meet the energy demands of the 21st Century. As the world population expands, and economic aspirations grow, alternatives must be found to the continued depletion of fossil fuels. While the creation of a space transportation infrastructure, lunar mining and processing facilities, and terrestrial power plants represents a formidable challenge, so did the Apollo Program. Unlike that undertaking, this new enterprise would require a much smaller investment of capital. But it would still rely on the talents of a highly motivated reservoir of people from throughout the world. Of course, the key ingredient to any great endeavor, as Wernher von Braun once observed, is "the will to do it."

Some day, a new generation of lunar travellers will walk in my footsteps and stand in the Valley of Taurus-Littrow. They will go to the Moon to further explore its wonders, and to employ its resources for the benefit of humankind. With the right combination of talent, vision, enthusiasm, and commitment, that day could come in the not-so-distant future. Based on our Apollo mission experience and the large-scale mining and refining expertise developed on Earth, commercial and profitable mining operations on the Moon could be established around 2015. This matches the estimates of the time required to develop and build the first commercial helium-3 reactors on Earth.

Humankind sought and attained galactic stature with the first lunar explorations between 1967 and 1973. During those momentous years, our species took its first clear steps of evolution into the Solar System and eventually into the galaxy. Now, as the Pueblo Indians relate the lesson of their ancestors, "We walk on the Earth, but we live in the sky."

I hope you enjoy Bill Mellberg's account of the early years of lunar exploration, and the lessons we learned during those first *Moon Missions*.

Harrison H. Schmitt   Albuquerque

About Harrison H. Schmitt

As the Lunar Module Pilot on the Apollo 17 mission in December, 1972, Harrison "Jack" Schmitt was the last man, and the *only* geologist, to set foot on the surface of the Moon. A member of the first class of scientist-astronauts selected by NASA in June, 1965, he worked with Dr. Eugene Shoemaker at the Astrogeology Branch of the U.S. Geological Survey (USGS) in Flagstaff, Arizona. Schmitt participated in the photographic and telescopic mapping of the Moon, and was among the USGS astrogeologists instructing NASA astronauts during their geological training trips.

After becoming an astronaut himself, Schmitt continued to play a major role in providing Apollo flight crews with detailed instruction in lunar navigation, geology, and feature recognition. He also helped integrate scientific activities into the Moon missions and participated in the analysis of lunar samples.

Following his trip to the Moon, Schmitt directed NASA's Energy Program Office in Washington, DC. In 1976, Schmitt was elected to a six-year term in the United States Senate from his home state of New Mexico. Today, he is a consultant on matters of science, technology, energy, and public policy.

The recipient of many awards and honors, Schmitt earned his Bachelor of Science degree from the California Institute of Technology (1957), studied under a Fulbright Fellowship at the University of Oslo in Norway (1957-1958), and received his Doctorate in Geology from Harvard University (1964). He currently serves as an adjunct professor of engineering at the University of Wisconsin in Madison.

Introduction

Many books have been written about the Apollo Program. Most of them have described the incredible experiences of the two dozen astronauts who flew to the Moon, and of the 12 men who walked on its surface. But few books have described what was actually found there.

The focus of *Moon Missions* is the Moon itself. The book answers five basic questions:

1. **Why did we go to the Moon?** Part One examines what we knew (and didn't know) about our nearest celestial neighbor prior to the Space Program. It discusses the visionaries who dreamed about a lunar journey and the political factors which eventually made that trip possible.

2. **How did we get there?** Part Two looks at the methods and machines men used to reach the Moon. The Moon missions—manned and unmanned, American and Soviet—are described in some detail with an emphasis on the purpose and results of each one. How were individual landing sites chosen? How did they differ from each other? How did the astronauts live and work on the lunar surface?

3. **What did we find there?** Part Three summarizes and synthesizes the scientific results of the Apollo Program and how they relate to our understanding of the rest of the Solar System, including our own planet. What did the lunar rocks tell us about the chemical composition and geologic history of the Moon?

4. **Was it worth the cost?** Part Three deals with the question: What, if any, were the economic, political, scientific, and social benefits of the 'Moon Race?'

5. **Should we go back?** Part Three discusses what justification can be offered for a return to the Moon. How can its unique energy resources be utilized here on Earth?

It has been nearly a quarter of a century since men walked on the Moon. An entire generation has grown up after Apollo—a generation that never knew the aura of mystery that enshrouded Earth's natural satellite or the excitement and drama that accompanied its exploration.

For those who did live through the Moon Race, three flights remain etched in memory: The 1968 Christmas Eve voyage of Apollo 8 around the Moon represented humanity's first break with Earth's gravity. The historic landing of Apollo 11 in 1969 marked our first steps on another world. And the dramatic rescue of Apollo 13 in 1970 symbolized the triumph of the human spirit over adversity.

Five other Apollo lunar landing missions returned a wealth of information about the Moon; the last three extended flights, in particular, being history's greatest voyages of discovery. These missions made interplanetary travel seem routine. They received scant acclaim at the time, and even less attention in the years that followed. In a sense, they are the forgotten stories of the Apollo era. Yet, they made enormous contributions to science and technology. I believe history will remember the Apollo Program long after many other 20th Century events have been forgotten.

As a young man, I was privileged to witness two remarkable moments in the pioneering years of lunar exploration. The first happened shortly after midnight on the morning of June 2, 1966. I was fourteen years old, and had to go to school the next morning, but my parents allowed me to stay up late to watch television as the unmanned Surveyor 1 spacecraft made America's *first* landing on the Moon. I will never forget that night because my father, Frank Mellberg, was the program manager responsible for Surveyor's television camera zoom lens. I stared in amazement at the pictures being broadcast from the lunar surface, and I was proud that my Dad had helped to make them possible.

The second remarkable moment came six and a half years later on December 7, 1972. It, too, was shortly after midnight. My father and I were standing near the giant Vehicle Assembly Building at the Kennedy Space Center, watching Cernan, Evans, and Schmitt leave Earth for America's last Apollo landing. It was an awesome sight that seized my imagination, and inspired this book.

The Moon beckons us to return. I believe we will resume our journey toward the stars, because, as Neil Armstrong said, "...there lies human destiny."

Bill Mellberg
Chicago
June 1997

The First Quarter Moon as Galileo must have seen it through his telescope in 1609. Clearly visible are the rugged regions along the terminator—the line dividing lunar night and day. Galileo was the first to see and record mountains and craters on the Moon's surface. (Copyright UC Regents Lick Obs)

Part One
The Old Moon

1 The Moon

Let us speak first of that surface of the Moon which faces us. I distinguish between two parts of this surface, a lighter and a darker. The lighter part seems to surround and to pervade the whole hemisphere, while the darker part discolors the Moon's surface like a kind of cloud, and makes it appear covered with spots. Galileo Galilei, 1610

As the Moon appeared to Galileo, so it appears to us today—a bright orb, serenely floating across the sky. The ancients noticed that the Moon's 29½-day cycle is made up of a series of phases—from a thin crescent, to a fully lit disk, to a crescent once again. Occasionally, they observed the temporary fading of the Full Moon's light during a lunar eclipse. Sometimes, the ancients watched in awe as the New Moon passed in front of the Sun during a solar eclipse. Even though these heavenly motions were readily apparent to all who viewed them, the nature of the Moon remained a mystery.

Was it just a light in the sky? Was it another world? Did life exist there? What caused the monthly phases? What were those dark spots that pocked its face? Many ancient cultures worshipped the mysterious Moon.

Around 130 B.C., the famous Greek astronomer, Hipparchus, carefully studied the Moon's motions and concluded that both it and Earth were spheres. He even measured the distance to the Moon, his mean value of 68 Earth radii being remarkably close to the actual figure of 60.27.

Still, little was known about the Moon until 1609, when Galileo pointed a new invention, the telescope, toward the sky. He discovered four new moons (now called the Galilean satellites) orbiting Jupiter, he noted the changing phases of Venus (similar to the Moon's), and he observed strange spots on the surface of the Sun. His sensational observations were published the following year in a booklet entitled *Sidereus Nuncius (The Starry Messenger)*.

Galileo's descriptions of the Moon astounded his contemporaries. He was the first person to see its countless craters and to correctly identify the Moon as another world. "I have been led to the opinion and conviction," he wrote, "that the surface of the Moon

is not smooth, uniform, and precisely spherical, as a great number of philosophers believe it (and other heavenly bodies) to be, but is uneven, rough, and full of cavities and
prominences, being not unlike the face of Earth, relieved by chains of mountains and
deep valleys."

As improved telescopes became available (Galileo's was no more powerful than a
modest pair of modern binoculars), observers discovered more detail on the lunar surface. In 1647, Johann Hewelke, a Polish astronomer better known as Hevelius, published
an atlas of the Moon, *Selenographia*. A few of the names Hewelke gave to lunar mountain
ranges are still in use, such as Alps, Apennines, and Caucasus. An Italian astronomer,
Giovanni Baptista Riccioli, devised the current system of lunar nomenclature in 1651.
Craters were named for famous scientists (e.g. Plato, Copernicus, Tycho) while the dark
areas, thought to be maria (seas) owing to their smooth appearance, received Latin names
such as Mare Tranquillitatis (Sea of Tranquillity), Sinus Iridum (Bay of Rainbows), and
Oceanus Procellarum (Ocean of Storms).

Succeeding generations of astronomers observed other features on the Moon's surface. Long, narrow, trench-like rimae (or rilles) crossed the landscape, as did curious
wrinkle ridges. The Straight Wall was a distinctive, 68 mile (110 km) long escarpment in
Mare Nubium (Sea of Clouds), apparently created when the lunar surface was displaced
by a fault. Scores of volcano-like domes were located in the maria. Enigmatic "rays"
(bright streaks) emanated from some craters, particularly during a Full Moon. The craters Tycho and Copernicus offered spectacular examples of this phenomenon.

All of these fascinating features are readily visible today, through even a small telescope. Other than rays, most lunar landmarks are best observed when the Moon is less
than full, especially when they are at or near the Moon's terminator—the slowly moving boundary between light and darkness. As this line, running from north to south,
marks either sunrise or sunset on the Moon (or on Earth), objects near it—such as
mountains—cast longer shadows and show greater relief. A good Moon map or atlas,
available at most libraries, can help identify specific craters and mountain ranges; and a
stable, well-mounted pair of binoculars can reveal many of the Moon's countless craters
and broader features, such as the lunar seas.

By the 19th century, astronomy had advanced sufficiently for a new picture of the
Moon to emerge. The dark "seas" were shown to be vast, dry plains, most likely lava-
covered. The lighter "highlands" were peppered with craters, some nearly 140 miles
(225 km) in diameter, and others less than a mile (1.6 km) wide. How these craters were
formed was a hotly-debated topic. Some experts argued that the lunar craters were large,
bowl-like, volcanic calderas similar to those found at the summits of some volcanoes on
Earth (e.g., Mauna Loa and Kilauea in Hawaii). Others countered that the craters were
blasted out by meteorites. But whether the Moon's craters were formed by volcanism or
by impacts, most astronomers agreed that the Moon was an arid, airless world, totally
inhospitable to life.

The Moon's rugged southern highlands (south is at the top in this photo) with the 140 mile (225 km) wide crater Clavius (near the horizon and pitted by several smaller craters) and the 53 mile (85 km) wide crater Tycho (just below and to the right of center, identifiable by its central peak). (Copyright UC Regents Lick Obs)

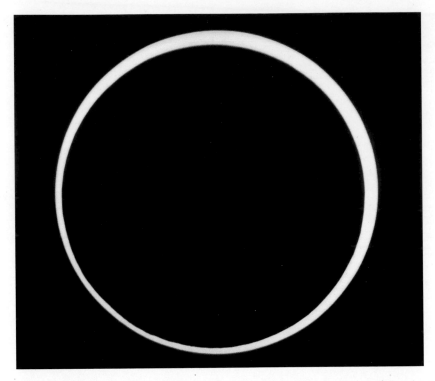

When the New Moon passes directly between Earth and the Sun, there is a solar eclipse. A total eclipse occurs when the Moon is at the low point of its orbit and appears large enough to completely cover the Sun. When the Moon is more distant, its apparent size is smaller than the Sun, resulting in an "annular eclipse," which occurred on May 10, 1994, when this photo was taken by the author.

Astronomers also determined the Moon's diameter (2,160 miles or 3,474 km), its average distance from Earth (239,000 miles or 384,400 km), and its surface gravity (one-sixth that of Earth). Because the Moon orbits Earth in an ellipse, its distance from us varies from 228,000 miles (364,800 km) to 252,000 miles (403,200 km). The Moon's surface area is about equal to the combined surface areas of North and South America. Noontime temperatures at the Moon's equator reach a blistering 243 degrees Fahrenheit (130 degrees Centigrade), and drop to a frigid minus 261 degrees Fahrenheit (minus 162 degrees Centigrade) at night. Lunar days and nights are each two Earth-weeks long.

The Moon always shows us the same face (the near side) because it rotates on its axis over the same time that it revolves around Earth, 27¼ days. This can be rather confusing, considering the fact that the time from one Full Moon to the next is 29½ days. The difference is explained by the complex relationship of Earth and the Moon to the Sun and stars, not the Moon's rotation on its axis and revolution around Earth. The point to remember is that we can only observe one of the Moon's hemispheres from Earth. The other hemisphere, the far side (*not* the dark side), remained unseen until space probes flew behind the Moon at the dawn of the Space Age.

In the late 18th century, two French scientists, Lagrange and Laplace, used Newton's law of gravity to show that a bulged or slightly egg-shaped Moon would keep its longest axis pointed toward Earth. This synchronous rotation is the result of Earth's strong gravitational attraction, and helps explain why we see only one side of the Moon. While Earth tugs at the Moon, the Moon's gravity has its own pull on Earth, causing the daily rise and fall of ocean tides.

By the time the 20th century dawned, the Moon's major physical characteristics had been determined. Yet many questions remained. Even the most powerful telescopes failed to resolve lunar features less than a quarter of a mile (0.4 km) wide. What was the Moon's origin? Did it form separately from Earth? Were valuable resources buried below its surface? Was that surface hard or soft? Was it covered by rock or dust? What formed the craters? To further unravel the Moon's mysteries, scientists would actually have to travel there and explore its surface. At the turn of the century, that journey seemed impossible.

2 The Visionaries

Born at Nantes, France, on February 8, 1828, Jules Verne was filled from an early age with wanderlust. Verne's father, a successful attorney, was determined to see his son follow in his footsteps. But law had no appeal to young Jules, although he did enjoy spending hour after hour in public libraries poring over classic works of fiction, travel books, and the latest scientific journals. Verne's stories, upon which rest his reputation as a literary giant, drew upon this storehouse of accumulated knowledge, as well as his many interests. His place of honor reflects not only his marvelous ability to weave a story, but his uncanny vision.

One of Verne's most famous works, *From the Earth to the Moon* (1865), was followed by a sequel, *Round the Moon* (1870). These stories were detailed accounts of a lunar voyage, based upon the latest astronomical and scientific discoveries. They also accurately predicted many of the problems inherent to spaceflight. Widely acclaimed, the books established Verne as one of history's most popular and enduring writers. More importantly, they inspired several generations of scientists and engineers to bring Verne's fantastic dreams to fruition. Indeed, the first Apollo flight to the Moon in 1968 closely paralleled Verne's prophetic vision.

Consider the following: In Verne's books, three men were shot to the Moon from Florida, in December, using a giant cannon. As their projectile circled the Moon, the crew made detailed observations of its surface. The "Moon men" returned to a splashdown in the Pacific, where they were recovered by the U.S. Navy. Then they enjoyed a

Jules Verne's From the Earth to the Moon, *first published in 1865, related the story of a voyage around the Moon. Here, an illustration from the book depicts the spaceship's passengers enjoying the effects of weightlessness.* (NASA)

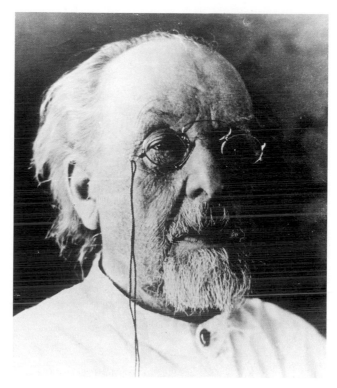

Konstantin Tsiolkovsky (1857-1935) is known in Russia as the "Father of Cosmonautics." He laid the theoretical foundations for multi-stage, liquid-fuel rockets, space stations, and interplanetary travel. (Tsiolkovsky Museum via V. Lytkin)

triumphant welcome home. In an extraordinary example of real life imitating fiction, a century later, NASA launched three men to the Moon from Florida, in December, using a giant rocket. As the Apollo 8 spacecraft orbited the Moon, the crew took detailed photos of its surface. The astronauts returned to a splashdown in the Pacific, where they were recovered by the U.S. Navy. They then enjoyed a triumphant welcome home.

A hundred years passed between Verne's fictional voyage and Apollo 8. *From the Earth to the Moon* and *Round the Moon* were read and reread by other visionaries in many different parts of the world. Three of them became space pioneers.

In Russia, an obscure school teacher and self-taught physicist, Konstantin Tsiolkovsky (1857–1935), living in the remote village of Kaluga, realized that Verne's giant cannon was quite impractical. No man could survive the acceleration forces when the cannon was fired. However, Tsiolkovsky recognized that a multi-stage, liquid-fueled rocket could accelerate more slowly and escape Earth's gravity, safely carrying humans to the Moon. Tsiolkovsky never built a rocket, but he was a prolific writer, and his books spelled out in great detail many basic theories of spaceflight. In Russia he has been called the "Father of Cosmonautics" because his works described not only rockets, but also spacesuits, space stations, lunar bases, and all the basic elements of interplanetary space travel. Several of his books, in particular *On the Moon* (1887), used science fiction to reach a general audience, but all of his books were based on scientific fact.

In America, a Clark University physics professor, Robert Goddard (1882–1945), was also inspired by Verne's tales. Like Tsiolkovsky, Goddard realized the potential of using liquid-fuel rockets to reach the Moon and published a series of papers on the subject. But Goddard went a step further, actually producing hardware. He launched the world's first successful liquid-fuel rocket from a farm in Auburn, Massachusetts, on March 16, 1926. It reached an altitude of just 41 feet (12.6 meters), but those were the first 41 feet to the Moon!

Dr. Robert Goddard (1882-1945) designed and flew the world's first liquid-fuel rocket, and patented much of the technology upon which modern rocketry is based. Here he examines the combustion chamber of one of his last and most advanced rockets in 1940 at his workshop in Roswell, New Mexico. (NASA via Goddard Space Flight Center)

Professor Hermann Oberth (1894-1989) with his former student, Dr. Wernher von Braun (1912-1977) at the American Rocket Society's "Old Timer" Dinner on October 10, 1961. Von Braun is holding the Society's Hermann Oberth Award, which he had just received for his many contributions to rocket engineering. (NASA via MSFC)

For the next 15 years, Goddard continued to develop his rockets and his ideas. Working in seclusion at a ranch near Roswell, New Mexico, and with the support of Charles Lindbergh and the Guggenheim Fund, his progress was slow but steady. Despite his modest resources, he perfected and patented much of the basic technology upon which modern rocketry is based. During his lifetime, Goddard's work was recognized, but largely unrewarded.

The last of the three early pioneers of space travel inspired by Jules Verne was a German, Hermann Oberth (1894–1989). Like Goddard, Oberth was a physics professor. His classic work, *The Rocket into Planetary Space,* published in 1923, was a brilliant paper which discussed both rocket design and space travel. When Goddard read it, he suspected that Oberth had borrowed some of his ideas, but it seems more likely that the two scientists had simply reached many of the same conclusions through their own independent studies. (Oberth had corresponded with both Tsiolkovsky and Goddard, but none of the three had ever met.)

Unlike either Tsiolkovsky or Goddard, Oberth welcomed publicity, and his work helped spawn Germany's Society for Space Travel. Oberth also served as a technical advisor to a German film, called *Frau im Mond (Lady in the Moon),* which described a lunar voyage. As with Verne's books, the film generated public interest in space travel, and rocketry became a popular sport in Germany during the 1920s and early 30s.

One wide-eyed young student who was attracted to the film and to Oberth's work was Wernher von Braun, the son of Baron Magnus von Braun, a Prussian aristocrat and pre-Hitler government official. Like his mentor, von Braun (1912-1977) dreamed of flights to the Moon. He had the benefit of Oberth's theoretical ideas, as well as his own keen engineering mind. Von Braun also had a charismatic personality, which, together with superb communication skills, made him an extraordinary manager. Under his leadership, Germany launched an A-4 rocket from Peenemünde on October 3, 1942. Reaching an altitude of nearly 60 miles (96 km), the A-4 was the single greatest advancement in the history of rocketry. Unfortunately, the A-4 was designed for a sinister purpose. Hitler dubbed it "V-2" — the second of his notorious vengeance weapons (the first being the V-1 Buzz Bomb). Thousands of V-2s were used to bomb targets in England and Belgium, during World War II.

After the war ended, von Braun and his brilliant team of rocket scientists and engineers moved to the United States, where they continued their research and development work. Germany's lead in rocketry was thus transferred to America, along with a supply of captured V-2s. Eventually, Oberth, too, crossed the Atlantic and joined his protégé in the United States.

Meanwhile, the Soviet Union was aggressively pursuing its own rocketry program. The Gas Dynamics Laboratory was founded in Moscow in 1921 to design and test a variety of rocket engines and propellants. Ten years later, the Group for the Study of Reactive Motion was formed. After the two organizations merged in 1933, they produced many of Soviet Russia's greatest rocket designers, including Sergei Korolev (1906-1966). Korolev was both a brilliant engineer and skillful manager — a Soviet "Wernher von Braun." Like von Braun, Korolev dreamed of Moon flights, and like the German, his genius was drafted into military service during World War II.

After the war, Korolev and his colleagues assimilated German V-2 technology and applied it to their own rocket designs. Like the Americans, the Soviets were keen on developing rockets as long-range weapons — intercontinental ballistic missiles, or ICBMs — which could deliver nuclear warheads to any point on the globe. The Cold War, that ideological and military struggle in which the United States and Soviet Union competed for world power and influence, fueled the development of these dreadful weapons. But it also gave impetus to the Space Race, and with it, the conquest of the Moon.

3 The Race is On!

On July 29, 1955 the White House proudly announced that the United States would launch "history's first artificial Earth satellite" some time between July, 1957 and December, 1958 as part of the International Geophysical Year. The International Geophysical Year would bring together some 5,000 scientists, representing over 40 countries, including the Soviet Union, in an intensive study of Earth and its environment from the ocean depths to the outer reaches of the atmosphere. The man-made moon would provide important new information about cosmic rays, meteoric particles, and solar radiation, as well as precise geodetic data regarding Earth's true shape. The Naval Research Laboratory was given the responsibility for Project Vanguard, as the satellite program was called, and the world looked forward to this exciting American effort.

On the night of October 4, 1957, a giant R-7, the seventh in a series of post-war Soviet rocket designs, roared off its launching pad at the Baikonur Cosmodrome, light

The first and second stages of the Soyuz launcher—used to send cosmonauts to Russia's Mir space station and shown here—are relatively unchanged from Korolev's original R-7 ICBM, used to launch Sputnik in 1957. At lift-off, no less than 32 rocket chambers fire in the core unit (second stage) and the four "strap-on" boosters of the first stage. (Novosti Press Agency)

ing up the sky over the Kazakhstan steppes. Atop the rocket, under a protective nose cone, was a highly-polished aluminum sphere, 22.8 inches (58 cm) in diameter, weighing 184.3 pounds (83.6 kg). A few minutes later, a steady "beep-beep-beep" was transmitted by the four long antennas affixed to the exterior of the Sputnik (Russian for Traveler). Those beeps were heard around the world, for the Soviet Union—*not* the United States—had launched the first artificial Earth satellite. The Space Age had arrived.

At the cosmodrome, an overjoyed Korolev spoke with pride and emotion to his jubilant rocket team. "Today the dream of the greatest representatives of mankind, including our famous scientist, Konstantin Tsiolkovsky, has come true. He predicted that man would not remain bound to our Earth forever. The Sputnik is the first confirmation of his forecast. The exploration of space has begun!"

The R-7 was actually the Soviet Union's first successful Intercontinental Ballistic Missile, or ICBM, as the long-range delivery systems for nuclear weapons are better known. Although it soon proved to be too big and unwieldy to be an effective ICBM, the tall tapered rocket was destined to become one of the world's most reliable space launchers. Korolev, the famous, though long anonymous, Chief Designer, was the man responsible for the R-7's overall design. The rocket's powerful RD-107 and RD-108 liquid-fuel engines, however, were developed by his brilliant colleague, Valentin Glushko (1906-1989). Both men had survived harsh imprisonment during Stalin's purges to pursue their dreams of space travel.

An RD-107 powered each of the four strap-on boosters that comprised the first stage of the R-7. In addition to its four main thrust chambers, the RD-107 employed two additional control chambers for steering. The RD-108 was a similar engine used in the second stage, but was equipped with four steering chambers. Both stages were powered at lift-off, meaning that no less than 32 rocket chambers were firing at the same time! The first stage boosters were simultaneously jettisoned when their propellants were exhausted, while the central (core) unit continued to burn until Sputnik achieved orbit.

On November 3, 1957, the Soviet Union stunned the world again. Aboard Sputnik-2, a dog, Laika, became the first living creature to travel into space. Although the Russians were very secretive about their intentions and their space hardware, they did reveal that Sputnik-2 was six times as heavy as its predecessor. The implication was clear; manned spaceflights could not be far behind. Much more disturbing was the fact that Sputnik-2 demonstrated the Soviet Union's ability to deliver heavy nuclear warheads against the United States or any other country it chose as a target.

The Americans made their attempt to put a satellite into orbit on December 6, 1957. In full view of the world, the Vanguard rocket lifted a few inches off the ground, settled back on its launching pad, and exploded in a ball of fire. It was an embarrassing episode which further eroded American prestige.

America's original "Mercury Seven" astronauts were introduced on April 9, 1961, just six months after the formation of NASA. Front row, left to right: Walter M. Schirra, Jr.; Donald K. "Deke" Slayton; John H. Glenn, Jr.; and M. Scott Carpenter. Back row, left to right: Alan B. Shepard, Jr.; Virgil I. "Gus" Grissom; and L. Gordon Cooper. Of these seven, only Alan Shepard, America's first man in space, would travel to the Moon. (NASA)

President Dwight Eisenhower turned to Dr. von Braun and his rocket team, now established at the Army Ballistic Missile Agency in Huntsville, Alabama, to salvage American prestige. After the launch of Sputnik-1 had been announced, von Braun had exclaimed, "We could have done it with our Redstone two years ago!" The Redstone rocket was a highly developed V-2. First flown in 1953, it was designed as a tactical weapon for the U.S. Army. But as early as 1955, von Braun had proposed using a modified Redstone, the Jupiter C (also known as Juno 1) to launch an artificial satellite. "The establishment of a man-made satellite, no matter how humble…would be a scientific achievement of tremendous impact," von Braun wrote at the time. "It would be a blow to U.S. prestige if we did not do it first."

The Vostok launcher, which propelled the first manned space flight, was erected at the Baikonur Cosmodrome on April 11, 1961. Based on Sergei Korolev's R-7 ICBM, it had a third stage added to enable it to send Vostok into orbit. Korolev's rockets were put together horizontally in the Assembly and Test Building (ATB), then transported by rail to the launch pad. (Novosti Press Agency)

Von Braun had been proven right! With the successful launch of Sputnik-1, his proposal was dusted off and reconsidered. After Sputnik-2, von Braun was given the go-ahead. He promised to put a satellite into orbit within 90 days. Eighty-five days later, he made good his promise. On the night of January 31, 1958, von Braun's Juno 1 lifted off from Cape Canaveral, putting the 31 pound (14 kg) Explorer 1 satellite into a 224 by 1,575 mile (360 by 2,534 km) orbit. Designed by the California Institute of Technology's Jet Propulsion Laboratory (JPL), Explorer 1 found a doughnut-shaped region of energetic atomic particles encircling Earth. Potentially hazardous to future astronauts, the Van Allen radiation belt was named for Dr. James Van Allen of the University of Iowa, who had built the instruments that detected and measured it. The belt was the biggest discovery of the International Geophysical Year.

The United States was now in the space race, but the Russians still held a commanding lead. "Let's not forget that our payload is only a fraction of what Sputnik-1 was, and a smaller fraction of Sputnik-2's," von Braun reminded his colleagues. "We are competing only in spirit with Sputnik-2 so far—not in hardware yet."

Two months later, the White House released a booklet, *Introduction to Outer Space,* which offered justification for a national space program. One of the ideas mentioned in the publication was a proposal to send an unmanned probe to the Moon to photograph its hidden side and to shed light on some of its other mysteries.

Toward that end, on July 29, 1958, President Eisenhower signed the National Aeronautics and Space Act, which created the National Aeronautics and Space Administration (NASA). Among other goals, NASA was given the responsibility of sending the first Americans into space. NASA's first mission was the Pioneer 1 probe, launched on October 11, 1958. Intended as a lunar orbiter, Pioneer 1 never achieved escape velocity (the speed needed to break free from Earth's gravitational pull), and plummeted into the Pacific Ocean. However, Pioneer 1 did reach an altitude of 70,574 miles (113,830 km), nearly a third of the distance to the Moon.

Over the next 60 days, two more Pioneer craft failed, but on January 2, 1959, the Soviet Union scored again. The Luna-1 probe was sent toward the Moon and passed

Yuri Gagarin, the first man in space, is seen in the cabin of his Vostok (Russian for East) spacecraft just before launch on April 12, 1961. His single orbit around Earth took 108 minutes. Unlike early American spacecraft, which splashed down in the ocean, Vostok touched down on land. However, Vostok cosmonauts were first ejected from their spacecraft at 22,000 feet (6,700 m) and landed, using their own parachutes. (Novosti Press Agency)

within 3,100 miles (5,000 km) of its target before going into orbit around the Sun, making Luna-1 the first man-made planet. Luna-1 was launched on a modified R-7 rocket, a third stage being added to reach escape velocity of 24,900 mph (39,840 kph). On September 12, 1959, Luna-2 became the first man-made object to reach the Moon, crashing into Mare Imbrium near the large crater, Archimedes.

The Soviets achieved an even greater success, when, on October 4, 1959, the second anniversary of Sputnik-1, they launched Luna-3, which flew behind the Moon and transmitted the first pictures of its far side. The crude, low-contrast, black and white photographs revealed some 70 percent of the Moon's hidden face, showing its rugged surface for the first time. Unlike the near side, there were few smooth mare regions, although one that was seen was called Mare Moscoviense (Moscow Sea). Two dark-floored craters were named for Konstantin Tsiolkovsky and Jules Verne, in tribute to the two visionaries who had pointed the way to the Moon. The stark differences between the near and far sides of the Moon demonstrated just how much remained to be discovered on the lunar surface.

In March, 1959, Pioneer 4 flew within 37,200 miles (60,000 km) of the Moon and became the second spacecraft to enter orbit around the Sun. It would be another three years (January 28, 1962) before America's Ranger 3 would pass within 23,000 miles (36,800 km) of the Moon. On April 26, 1962, Ranger 4 crashed into the Moon's far side.

The Soviets did not enjoy any more great successes beyond Earth orbit until July of 1965, when Zond-3 finally surpassed Luna-3's mission by sending back photographs of previously unseen regions of the Moon's far side.

While both countries continued their unmanned exploration of the Moon, they were also busy preparing to send men into space. NASA introduced the Mercury Seven astronauts to the world on April 9, 1959. One of them was to fly the first sub-orbital mission within a year, and another was to make the first orbital flight by the fall of 1960. It was an ambitious, if not impossible, schedule. But NASA's public pronouncements encouraged the Soviet leader, Nikita Khrushchev, to drive Korolev's team even harder. Space firsts held enormous propaganda value, and Khrushchev understood their impact on the world stage.

By early 1960, the first 20 Soviet cosmonauts had been selected and were in training. Like their American counterparts, they were experienced military jet pilots. And, like the astronauts, each cosmonaut was anxious to become the first man in space. The race was heating up.

The Soviet Chief Designer, Korolev, had an advantage over his American competitors—a more powerful rocket, capable of carrying larger payloads into space. Korolev called on the R-7 once again. This time, the third stage used on the Luna missions was modified to boost Korolev's Vostok (Russian for East) spaceship. Work began on the spacecraft itself in 1959. The ship was divided into two parts, a descent module, which would carry the cosmonaut, and an instrument compartment, which would house oxygen bottles, electronic systems, batteries, radios, orientation thrusters, and a retro-rocket. After considering several shapes for the descent module, a sphere was chosen as the simplest solution to multiple design challenges. By the spring of 1961, several unmanned Vostoks had been tested in space, two of them carrying dogs (the Americans were using monkeys and chimpanzees to test their Mercury spacecraft).

The United States fell behind in its ambitious schedule to launch the first man into space by 1960. The initial, suborbital hops were to be accomplished with von Braun's reliable Redstone missiles, while the orbital trips would be made using the new, but not-so-reliable Atlas ICBM. Problems with both the rockets and the spacecraft which would carry the astronauts delayed Project Mercury by weeks, then months. With an astronaut's life at stake, Dr. von Braun urged caution. He added an extra Redstone test flight in March 1961, and postponed the first manned mission until late April or May.

On April 11, 1961, Korolev moved a Vostok rocket to its launching pad at the Baikonur Cosmodrome. Just three days earlier, a State Commission had chosen a 27-year old Air Force major, Yuri Gagarin, to ride the Vostok-1 into space. On April 12, 1961, Gagarin, dressed in a bright orange spacesuit topped off with a white helmet bearing the red letters CCCP (USSR), arrived at the launch pad by bus, where he was embraced by Korolev and cheered by his comrades. Taking an elevator to the top of the rocket, Gagarin climbed into the sphere and prepared to make history.

At 9:07 A.M. Moscow time the rocket roared to life. Gagarin shouted, "We're off!" A few minutes later, he was in orbit. Radio Moscow broadcast the news that "the world's first manned spaceship" had just been launched, and that the cosmonaut was "feeling well." By 10:55 A.M., Gagarin was safely back on the ground. He had circumnavigated the globe in just 108 minutes! He was quickly dubbed "the Columbus of Space."

Premier Khrushchev gave Gagarin a hero's welcome in Moscow. Khrushchev reveled in the Soviet Union's latest triumph over the Americans. "Let the capitalist countries catch up with our country which has blazed the trail into outer space," he boasted. "The dream of conquering outer space is indeed the greatest of man's dreams. We are proud that Soviet people have made this dream, this fairy tale come true . . . This exploit marks a new upsurge of our nation in its onward movement toward communism." According to Khrushchev, the Soviet Union had time and again demonstrated the superiority of its socialist science and technology over that of the capitalist United States. His public taunts inspired a bold response.

John F. Kennedy, America's youthful new president, had been in office for less than ninety days when he was faced with the Soviet propaganda nightmare. His problems with the Soviets were soon compounded by the disastrous invasion of Cuba at the Bay of Pigs on April 15. Not even the highly successful 15-minute flight of Astronaut Alan B. Shepard aboard the Mercury spacecraft, *Freedom 7,* on May 5, 1961, could ease the humiliation of Sputnik, Luna, and Vostok. Once again, the United States had answered

Alan Shepard became America's first man in space with a 15-minute sub-orbital flight on May 5, 1961. His Freedom 7 Mercury *spacecraft was launched from Cape Canaveral atop a modified Redstone rocket designed by Wernher von Braun's rocket team.* (NASA)

Russia's challenge with too little, too late. With American prestige on the line, Kennedy knew he had to do something dramatic to restore his nation's self-image and global reputation.

In a memo to the chairman of the Space Council, Vice President Lyndon Johnson, Kennedy asked, "Do we have a chance of beating the Soviets by putting a laboratory in space, or by a trip around the Moon, or by a rocket to land on the Moon, or by a rocket to go to the Moon and back with a man?"

Johnson consulted with a number of experts, including NASA's astute new administrator, James E. Webb, and Webb's highly respected deputy, Hugh L. Dryden. Johnson also called in Wernher von Braun. "We have an excellent chance of beating the Soviets to the first landing of a crew on the Moon," von Braun assured Johnson, so the Vice President made the case for sending a man to the Moon.

President Kennedy accepted the challenge. "Now is the time to take longer strides," he told a joint session of Congress on May 25, 1961. "Time for a great new American enterprise—time for this nation to take a clearly leading role in space achievement, which, in many ways, may hold the key to our future on Earth… I believe that this nation should commit itself to achieving the goal, before this decade is out, of landing a man on the Moon and returning him safely to the Earth." Kennedy stressed that "No single space project in this period will be more impressive to mankind, or more important for the long-range exploration of space. And none will be so difficult or expensive to accomplish."

Speaking in Texas the following year, Kennedy said, "We choose to go to the Moon in this decade and do the other things, not because they are easy, but because they are hard, because that goal will serve to organize and measure the best of our energies and skills, because that challenge is one we are unwilling to postpone, and one which we intend to win…" The Space Race had become a Race to the Moon!

The thunderous roar of the Saturn V could be heard — and felt — for miles. Here Apollo 11 blasts off on July 16, 1969. Just four days later, Neil Armstrong and Buzz Aldrin became the first men to walk on the Moon. (NASA via MSFC)

Part Two

The Moon Missions

4 Lunar Flight Plan

Having decided to send men to the Moon, President Kennedy charged NASA with designing a flight plan and building the hardware that would get them there. The project had a name, Apollo (after the Greek god of prophecy, music, light, and progress), but lacked a *modus operandi* for atcually reaching the Moon.

Wernher von Braun had proposed one method in a popular series of articles he wrote for *Collier's* magazine between 1952 and 1954. Von Braun's concept called for the development of a delta-winged, reusable space shuttle, and a large space station that would orbit Earth at 1,000 miles (1,600 km). The space station would serve as the jump-off point for a modified shuttle craft which would make the "final ascent to the Moon." Von Braun reintroduced this concept in the 1954 *Tomorrowland* television programs produced by Walt Disney. The attraction of building an elaborate infrastructure was that it could support a wide variety of space activities on a long-term basis, including an eventual trip to Mars.

While building a space station and shuttle before going to the Moon made a great deal of sense, NASA had neither enough time nor money. Three other methods were being considered. The first was called the Direct Ascent Mode and would have used a monstrous Nova rocket to launch Apollo directly to the Moon. The second, termed Earth Orbital Rendezvous, would have employed two giant Saturn V rockets, one to launch Apollo into Earth orbit where it would be met by a tanker sent up by the other Saturn. Apollo would then top off its tanks for the voyage to the Moon and back. The third method was called Lunar Orbit Rendezvous. In this scheme, a single Saturn V would be used to send a two-part Apollo spacecraft to the Moon, consisting of a Command Module attached to a Service Module, and a Lunar Excursion Module. Three men would fly to the Moon inside the Command/Service Module. Once in lunar orbit, two of them would descend to the surface in the Lunar Excursion Module (later referred to simply as Lunar Module or LM). Then they would rejoin their companion in the Command/Service Module for the return to Earth.

The Direct Ascent Mode and Earth Orbital Rendezvous plan would each land the entire spacecraft (and all three men) on the lunar surface. That same spacecraft would then make the return journey to Earth. Being built to land on Earth, the vehicle would be quite large and very heavy, requiring more fuel and bigger rockets. On July 11, 1962, NASA Administrator, James E. Webb, called a press conference in Washington to announce that Lunar Orbit Rendezvous had been chosen as the best method to reach the Moon. He cited four main reasons:

1. **It provided a higher probability of mission success** than the other modes with the same degree of safety.
2. **It promised mission success in less time** than the other modes, therefore making a Moon landing possible by 1970.
3. **It would cost an estimated 10 to 15 percent less** than the other modes.
4. **It would require the least amount of technical development** while still advancing the nation's technology.

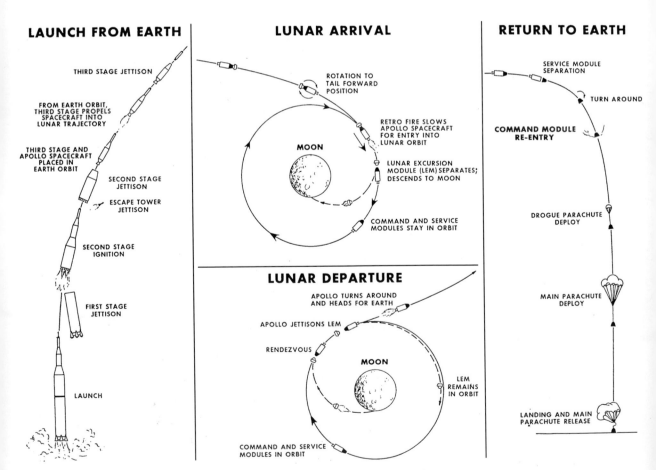

The major steps in the Apollo flight plan, known as Lunar Orbit Rendezvous, are shown in this illustration. (NASA)

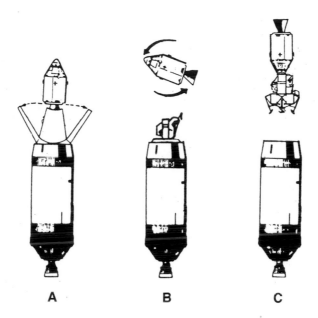

A   B   C

Shortly after leaving Earth orbit on their way to the Moon, the Apollo astronauts would perform a cosmic ballet called "Transposition and Docking." This was necessary in order to extract the Lunar Module from its protective cocoon atop the S-IVB third stage. First, the Command/Service Module pulled ahead of the vehicle, and the four petal-like pieces of the Spacecraft Launch Adapter separated and were allowed to drift off (A). Then the Command/Service Module turned around 180 degrees so that its nose was pointed toward the Lunar Module, which was still attached to the S-IVB (B). The Command Module Pilot fired thrusters to gently dock with the Lunar Module. Next, 12 capture latches were activated, firmly holding the two craft together. Finally, using reverse thrusters, the Command Module Pilot extracted the Lunar Module from the S-IVB (C). (NASA)

At first glance, Lunar Orbit Rendezvous appeared to be the most complex mode. But as early as 1916, a Russian rocket pioneer, Yuri Kondratyuk (1897-1942), had explained why such an approach would be the most energy-efficient: landing a lighter craft on the Moon meant burning less fuel. This same idea was also described by von Braun in his book, *The Mars Project* (1953). However, the main proponent of Lunar Orbit Rendezvous for Apollo was a NASA engineer, John Houbolt. He faced a great deal of opposition because Lunar Orbit Rendezvous entailed complicated flight maneuvers, requiring sophisticated computer programs. Houbolt countered that Lunar Orbit Rendezvous divided a huge problem into simple parts: one module designed to travel between Earth and the Moon, and the other built to land and takeoff on the Moon, permitting the engineers to focus their attention on specific problems. This rationale finally convinced NASA management to accept Lunar Orbit Rendezvous.

As adopted, the plan called for the launch of an Apollo spacecraft atop a three-stage Saturn V Moon Rocket. The S-IVB third stage of the Saturn V would also serve as the second stage of the smaller Saturn IB rocket, which would be used to test the Apollo modules in Earth orbit before attempting trips to the Moon. The S-IVB stage, like the S-II second stage on the Saturn V, would make use of the highly-efficient liquid hydrogen first proposed as a rocket fuel by Tsiolkovsky in 1903.

On the lunar voyage, after the Saturn V's first and second stages had finished firing and had dropped away, the S-IVB would place Apollo into a parking orbit around Earth. Once the spacecraft's systems were thoroughly checked, the S-IVB would be ignited once again to allow the spacecraft to reach escape velocity, sending Apollo to the Moon. En route, the Command/Service Module (CSM) would separate from the S-IVB, turn around, dock with the Lunar Module (LM) which had been stored in a protective launch adapter below the CSM, and pull it away from the third stage. The Service Module (SM) engine would brake the combined spacecraft as it neared the Moon, putting Apollo into lunar orbit. One astronaut would remain in orbit in the Command Module (CM) portion of the CSM while his two companions entered the LM. Using the LM's Descent Stage engine, they would land on the Moon, where they would explore its surface, gather rock specimens, and leave behind scientific experiments.

For the return to the orbiting Apollo CSM, the LM's Ascent Stage engine would be ignited, using the Descent Stage as a launching platform. After climbing into lunar orbit, the Ascent Stage would rendezvous with the CSM and dock with the CM. Then the lunar explorers would float back into the CM, transfer their precious cargo, and separate from the LM, leaving it in lunar orbit. The reunited astronauts would once more fire the SM engine. This trans-earth injection burn would propel them homeward. Just prior to entering Earth's atmosphere, the astronauts would jettison the SM. The heavily-shielded, blunt end of the CM would then protect the spacecraft and its occupants during a fiery reentry. Parachutes would gently lower Apollo to an ocean splashdown, ending a historic, eight-day lunar odyssey.

NASA had its plan.

It was time to build the hardware.

5 The Spaceships

Plans for Project Apollo were first announced during a NASA-Industry Conference on July 28, 1960. Apollo was originally conceived as an advanced three-man spacecraft designed for both Earth orbital and circumlunar missions (flights around the Moon). NASA managers noted that Apollo could also be used to directly support future Moon landings. Ten months after the conference, President Kennedy made a manned lunar

COMMAND MODULE

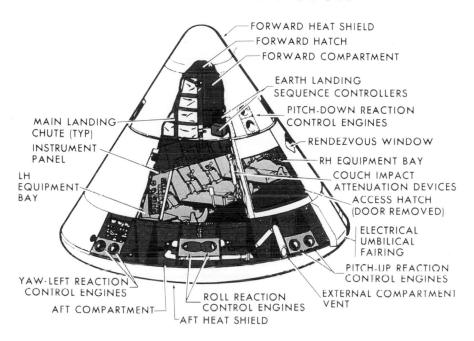

This drawing of the Apollo spacecraft Command Module shows its major systems. The astronauts sat in three couches during launch and re-entry. A large instrument panel was mounted directly over their heads. The Command Module was also equipped with control thrusters and parachutes, as well as a forward hatch and docking collar at the nose. (NASA)

landing a national goal. With the adoption of Lunar Orbit Rendezvous for the flight plan, the basic Apollo spacecraft as previously envisioned would be used as a mother ship, while the lunar lander would be developed separately. North American Aviation, Inc. (later Rockwell International) of California was given a contract to produce the Command and Service Modules in November, 1961. One year later, Grumman Aircraft Engineering Corporation of New York was chosen to build the Lunar Excursion Module.

At launch, the Apollo spacecraft was comprised of the Command and Service Modules, the Lunar Module, a Spacecraft–Lunar Module Adapter, and a Launch Escape System. The Spacecraft–Lunar Module Adapter protected the LM, which was housed during ascent between the S-IVB stage and the Service Module (SM). The Launch Escape System was perched atop the Command Module (CM) using an open-frame tower structure, and was designed to propel the CM and its crew to safety in the event of an aborted launch. (The three main solid-propellant motors of the Launch Escape System produced nearly twice as much thrust as the Redstone that had carried Alan Shepard into space aboard *Freedom 7*.) The CM would be shielded during the Saturn V's ascent by a fiberglass Boost Protective Cover which was carried away when the Launch Escape System was jettisoned (about three and one-half minutes into the flight).

From the base of the Spacecraft-Lunar Module Adapter to the tip of the Launch Escape System, the Apollo spacecraft stood 82 feet (25 m) tall. The cone-shaped CM was 11 feet 5 inches (3.5 m) high and 12 feet 10 inches (3.9 m) wide at its base. It provided the control and living quarters for the three-man crew, and brought them home at the end of their mission. The CM was encased in heat shields to protect the crew from the tremendous frictional heat that would be generated when the spacecraft plunged back

SERVICE MODULE

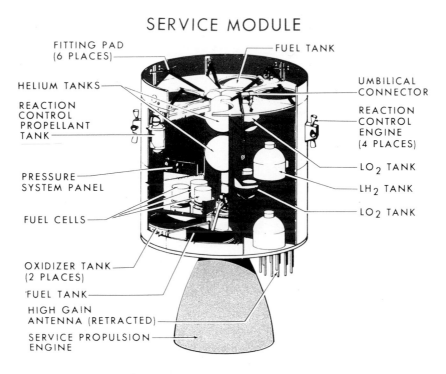

The Apollo Service Module shown here was mounted to the Command Module until just prior to re-entry into Earth's atmosphere. The Service Module provided oxygen, water, and electricity for the astronauts. It also housed the fuel and oxidizer tanks for the 20,500 pound (9,300 kg) thrust Service Propulsion Engine that propelled the spacecraft into and out of lunar orbit. Four Reaction Control System quads mounted around the circumference of the Service Module provided steering for Apollo. (NASA)

into Earth's atmosphere. Twelve reaction control engines provided directional control during reentry. Huge landing parachutes were stowed in the nose.

The astronauts sat in three padded couches, mounted side-by-side beneath a large instrument panel. The crew entered the spacecraft through a side hatch, and a forward tunnel gave them direct access to the docked Lunar Module. Stowage lockers for food and other supplies, including sleeping bags for two of the astronauts, were located under the crew couches. The center couch could be folded and stowed to give the cabin more usable space. The CM was equipped with two side windows, a hatch window, and two forward-facing windows that allowed views of the rendezvous and docking operations. During a typical lunar Mission, crew members would spend most of their time in the CM.

A docking probe in the nose of the CM captured a drogue assembly that was mounted atop the Lunar Module (LM). The funnel-shaped drogue was specifically designed as a

LUNAR EXCURSION MODULE

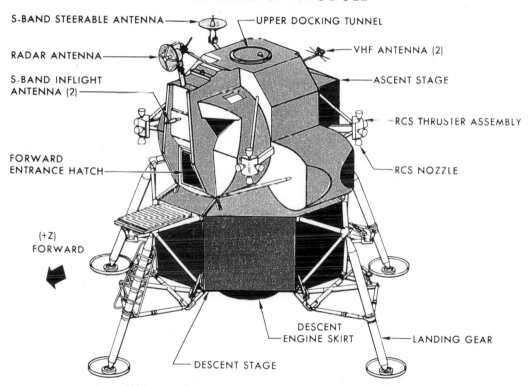

The main components of the Lunar Excursion Module (later called Lunar Module) are shown here. For Moon landings, two astronauts stood side-by-side in the Ascent Stage, peering through large triangular windows which allowed forward and downward views. On the Moon, they exited through the forward hatch and climbed down to the surface via a ladder mounted on the forward landing gear. (NASA)

stabilizer and receptacle for the pin-like probe. Twelve latches surrounding the CM 's docking ring provided a firm connection between the two vehicles—the probe and drogue being removed from the tunnel after docking.

Attached to the CM was the cylindrical SM with its large, 20,500 pound thrust (9,300 kg) Service Propulsion Engine that was used for mid-course corrections, lunar orbit insertion, and trans-earth injection. Mounted around the outside of the SM were four Reaction Control Systems quads that gave the CSM in-flight directional control, especially during rendezvous and docking maneuvers. Inside the SM were fuel and oxidizer tanks for the Service Propulsion Engine and Reaction Control System engines, as well as spacecraft oxygen, hydrogen, fuel cells (which generated both electrical power and water), environmental control equipment, and related systems. A steerable S-band High-Gain antenna was mounted at the rear of the SM for communications with Mission Control. During the last three Apollo lunar missions, additional hydrogen and oxygen tanks gave the spacecraft increased endurance for longer flights. A Scientific Instrument Module bay could accommodate a variety of high-resolution cameras and experiments which would be operated while in lunar orbit.

The two-stage LM was 22 feet 11 inches (7 m) tall and was designed specifically for landing on the Moon. Built to stand in the Moon's one-sixth gravity, the LM would collapse of its own 33,205 pound (125,059 kg) weight on Earth. The upper Ascent Stage housed a crew compartment for the two astronauts, who stood during landing and takeoff. Omitting seats saved weight, and standing gave the crew a better downward view of the lunar surface through the two large triangular windows which would be used during the critical landing phase of their flight. A small, rectangular window, located above the commander's head, would be used during rendezvous and docking. A tunnel and hatch at the top of the LM provided access to the CM, while a forward hatch between the astronauts gave them access to a "front porch" and ladder on which they would climb down to the lunar surface.

While inside the LM, the astronauts faced the forward instrument panels. Equipment for the Moon walks was located at the rear of the cabin, as was the top of a 3,500 pound (1,589 kg) thrust Ascent Engine. The Ascent Stage also carried water, oxygen, batteries, and fuel. Communication antennas and four Reaction Control System quads similar to those on the SM were mounted on the exterior to provide control during all phases of flight.

The lower Descent Stage housed a 9,870 pound (4,381 kg) thrust Descent Engine for landing the LM as well as fuel and oxidizer tanks, water, batteries, and scientific experiments which would be deployed on the lunar surface. Four landing gear legs, equipped with shock absorbers and footpads, were mounted around the Descent Stage. Sheets of black and gold thermal insulation covered the LM.

The weight of the Apollo spacecraft at launch was approximately 110,000 pounds (49,900 kg). Of that total, only the 12,250 pound (5,556 kg) CM would return to Earth, and—unlike today's Space Shuttle—it was not reusable. Viewed from this perspective, the Apollo spacecraft was not very practical. But it was designed with one overriding goal—to carry men to the Moon and back. As such, the Apollo Moonship was the most complex and sophisticated machine that man's ingenuity had yet conceived.

6 The Moon Rocket

Wernher von Braun was fond of drawing comparisons to illustrate the tremendous size and power of the Saturn V Moon Rocket. Towering 363 feet (110.6 m) high with the Apollo spacecraft mounted on top, it was, as von Braun noted, "taller than the Statue of Liberty." Each of the five F-I engines in the S-IC first stage produced 1.5 million pounds (680,400 kg) of thrust for a total of 7.5 million pounds (3,402,000 kg) at lift-off. That was enough power, as von Braun said, "to send one fully-loaded DC-3 airliner all the way around the Sun." And he quipped, "It could probably boost a Chevrolet clear out of the Solar System…there's no telling how far it would send a Volkswagen!" Weighing over six million pounds (2,721,600 kg), fully loaded, the Saturn V "would balance the scales with a light naval cruiser."

The Saturn V's first stage engines consumed 1,250 gallons (4,731 liters) of kerosene fuel and 2,083 gallons (7,884 liters) of liquid oxygen each second! (Since fire cannot burn without oxygen, and there's no oxygen in space, rockets must carry their own supply of oxidizer.)

By any measure, the Moon Rocket was a technological marvel and an engineering masterpiece. Developing the Saturn V had proved to be a very difficult task. It was the largest *operational* launch vehicle ever flown, representing a quantum leap in modern rocketry.

The gap between this monster and the Redstone missile, which produced a mere 78,000 pounds (35,380 kg) of thrust at lift-off—about one-hundredth of the Saturn V's first stage—was bridged by another pair of Saturns, the Saturn I and the more advanced Saturn IB. Initially designed by von Braun's rocket team in 1958, the Saturn I, America's first heavy lift launcher, was produced by clustering eight of the H-1 engines used in the Jupiter and Thor missiles to create a first stage that produced 1.5 million pounds (675,000 kg) of thrust. First launched on October 27, 1961, the Saturn I was at the time the world's mightiest rocket. But it still was only one-fifth as powerful as its descendent, the giant Saturn V.

The ten Saturn I flights proved the basic concept and paved the way for the Saturn IB. First flown on February 26, 1966, the Saturn IB was slightly more powerful, its first stage engines producing 1.6 million pounds (725,760 kg) of thrust. Nine Saturn IBs were eventually flown, five of them with Apollo and Skylab astronauts on board. The Saturn IBs were also used to test unmanned Command/Service Modules and Lunar Modules in Earth orbit. The single J-2 engine powering the S-IVB second stage produced 200,000 pounds (90,720 kg) of thrust. Standing 225 feet (68.6 m) tall, the Saturn IB was an essential stepping stone to the Moon.

But it was the giant Saturn V Moon Rocket that would actually carry the first lunar explorers on their epic voyage. The go-ahead to start work on the project was given in

The Saturn IB rocket was used for testing the Apollo spacecraft in Earth orbit. Its first stage was made up of eight H-1 engines. This particular Saturn IB (SA-205) launched the first manned Apollo flight (Apollo 7) on October 11, 1968. (NASA)

1962. The challenge that von Braun's team undertook was shared by the best industrial engineering talent in America. Boeing built the S-IC first stage; North American, the S-II second stage powered by five J-2 engines; and Douglas, the S-IVB third stage. The F-1 and J-2 engines were produced by the Rocketdyne Division of North American Aviation.

Mounted to the top of the S-IVB (on both the Saturn IB and Saturn V), just below the spacecraft, was a three foot (0.9 m) tall Instrument Unit which monitored the vehicle's performance from lift-off to orbit. Equipped with guidance and control equipment, it made use of computers to keep the rocket on course. If one engine shut down prematurely, the Instrument Unit kept the other engines burning long enough to compensate for the loss of thrust. This enabled the spacecraft to achieve a precise orbit at just the right speed and altitude. Designed by von Braun's group, the Instrument Unit was manufactured by IBM.

The Saturn V's mission was short-lived, lasting just 11 minutes from launch to orbit. It began at the moment of ignition, some seven seconds before lift-off. The first stage engines came to life with a roar, building thrust until the hold-down arms were released, permitting the giant vehicle to slowly begin its ascent. Two and one-half minutes later, the S-IC had carried the Moon Rocket to a height of 36 miles (57.6 km), and to a speed of about 6,000 mph (9,600 kph). As the first stage dropped away, five J-2 engines in the second stage were ignited. Making use of supercold, highly-efficient liquid hydrogen fuel, the S-II stage burned for six and one-half minutes and carried the vehicle to an altitude of 108 miles (173 km) and a speed close to orbital velocity, roughly 17,500 mph (28,000 kph).

Finally, the S-IVB third stage was lit for two minutes to place itself and the spacecraft into the desired parking orbit around Earth. Then, at just the right moment, the S-IVB was re-ignited to put Apollo on course to the Moon. During its six-minute burn, the S-IVB reached the escape velocity of 24,900 mph (39,840 kph), and the astronauts would watch Earth rapidly falling away behind them. After the Command Module docked with the Lunar Module, the spacecraft separated from the third stage, which went into orbit around the Sun. (On later missions the third stage was sent crashing into the Moon where lunar seismometers recorded the effect of its impact.)

The success of the first Saturn V flight on November 9, 1967, was a milestone in the Apollo Program. It was also a magnificent sight to behold—a towering white rocket with its black tracking stripes, huge red letters, and giant American Flags, climbing into the Florida sky on a long orange tongue of flame. The thunder from its first stage engines could be heard and felt for miles.

The United States finally had a rocket that could send men to the Moon.

7 Moonport

The gargantuan facilities necessary to assemble and launch the Saturn V rocket were as impressive as the Moon Rocket itself. Jules Verne fired his fictional astronauts to the Moon from a giant cannon located at a site near Tampa, Florida. The real Moon rockets were launched from a triangular-shaped peninsula on the eastern shore of Florida, called Cape Canaveral. In 1949, President Harry S. Truman had approved the construction of a long-range missile proving ground at "the Cape." The site was chosen in part because its position on the Atlantic coast would minimize the threat to populated areas from run-away rockets. The first missile launch at the Cape took place on July 24, 1950, with the firing of a captured V-2, part of America's booty from World War II. Over the next decade, hundreds of rocket launchings would occur from the Cape, most of them testing Air Force ICBMs.

In August 1961, NASA purchased a large tract of land on Merritt Island, north of and adjacent to the Cape. After the assassination of President Kennedy on November 22,

Moonport. *Launch Complex 39 was comprised of several structures, including the 52 story Vehicle Assembly Building with its high and low bay areas and giant doorways (outlined in black at the front of the building) located to the right of center; three Mobile Launchers, which can be seen parked on the left; two huge Crawler-Transporters, just below the Mobile Launchers; the Crawlerway, leading away from the Vehicle Assembly Building at the right; and two launch pads which are visible as white smudges on the horizon. This view looks east toward the Atlantic Ocean.* (NASA via Kennedy Space Center)

One of the two monstrous Crawler-Transporters. The deck on top is as large as a baseball diamond (note the temporary construction trailer parked there, as well as the man standing in front for a sense of size). One of two driving cabs can be seen at the lower left hand corner. (NASA via Kennedy Space Center)

1963, the spaceport that was already taking shape on the island was named the John F. Kennedy Space Center. It was also known as Launch Complex 39. (Cape Canaveral was called Cape Kennedy from 1963 to 1973.)

The major components of Launch Complex 39 were:

- **The Vehicle Assembly Building**, where the Apollo/Saturn V vehicle was assembled and readied for flight.
- **The Mobile Launcher,** upon which the vehicle was erected for checkout, transfer, and launch.
- **The Crawler-Transporter,** which transferred the vehicle to one of two launch sites.
- **The Crawlerway,** upon which the Crawler-Transporter traveled to the launch site.
- **The Mobile Servicing Structure,** which provided external access to the Apollo/Saturn V vehicle at the launch site.
- **Twin launch pads,** designated 39A and 39B.

Each of the components was built on a gigantic scale. The Vehicle Assembly Building, for instance, included a high-bay area, 525 feet (160 m) tall, and a low-bay area, 210 feet (64 m) tall. Together, in terms of volume, the two areas constituted the world's largest building, and allowed engineers to checkout and assemble flight-ready Apollo/

The Vehicle Assembly Building (VAB) enabled erection of the giant Apollo/Saturn V in an enclosed environment, protected from the elements. Here the Apollo 8 spacecraft is hoisted high above the floor of the VAB for mating to the Saturn V Moon Rocket. The entire vehicle was assembled on a Mobile Launcher, which was then carried to the launch pad by a Crawler-Transporter. (NASA)

Saturn V vehicles in an environment protected from the threat of tropical hurricanes or thunderstorms. Rocket stages arrived in the low-bay area where they were checked prior to assembly. Then the vehicle was carefully stacked on a Mobile Launcher in one of four high-bay assembly areas using 141 lifting devices, ranging from small hoists to huge bridge cranes.

Three Mobile Launchers were built, each one made up of a two-story steel base, a 380 foot (116 m) tall umbilical tower, and nine swing arms that provided access to the rocket and spacecraft prior to launch.

A Saturn V rocket with the Apollo 9 spacecraft mounted on top is rolled out of the Vehicle Assembly Building's High Bay 3 on its way to Launch Pad A. The Crawler-Transporter moved the 12,000,000 pound (5,443,200 kg) load at the snail's pace of one mile (1.6 km) an hour. Note the service arms reaching out from the Mobile Launcher to the Apollo / Saturn V. The topmost arm (Number 9) includes the "White Room" which provided access to the Apollo Command Module. (NASA)

Getting the 11,500,000 pound (5,216,000 kg) Mobile Launcher and Apollo/Saturn V vehicle from the Vehicle Assembly Building to the launch pad, 3.5 miles (5.6 km) away, posed quite a challenge. Sandy soil precluded rail transport, and barges were also impractical. NASA engineers came up with an ingenious solution, modifying the design of the giant crawlers that had been developed for strip mining operations. Two 6,000,000 pound (2,722,000 kg) Crawler-Transporters, each measuring 131 by 114 feet (39.9 by 34.7 m), were built. These monsters moved at one mph (1.6 kph) on four double-tracked crawlers, each track measuring 10 feet (3.0 m) high by 40 feet (12.2 m) long. Each crawler cleat weighed around 2,000 pounds (907 kg). From his position in one of two driving cabs located diagonally at each end of the Crawler-Transporter, the operator navigated turns and kept the huge load level. Sixteen hydraulic jacks lifted the Mobile Launcher from its pedestals inside the Vehicle Assembly Building, then they lowered the enormous load onto another set of matching pedestals after the journey to the launch site.

A two-lane, 8 foot (2.4 m) thick, crushed-rock Crawlerway ran from each of the Vehicle Assembly Building's four high-bay areas to the launch pads. The Crawler-Transporter's wide tracks evenly distributed its enormous load over this surface. In operation, the Crawler slipped under the Mobile Launcher, inside the Vehicle Assembly Building, entering and exiting through huge 456 foot (139 m) high openings. Each of these openings was shaped like an inverted "T." The upper sections were 76 feet (23 m) wide. The lower sections were 152 feet (46.3 m) wide, leaving room for the Crawler-Transporters to pass through. Together, these upper and lower sections created the world's largest doorways.

Adjacent to the Vehicle Assembly Building was a four-story Launch Control Center that housed monitoring and control equipment for checkout and launch operations. The actual countdown and launch was conducted from one of four firing rooms located inside this building.

After completing its slow, 3½ hour journey to the massive, polygon-shaped, concrete launch pad, the Crawler-Transporter lowered the Mobile Launcher onto steel support fittings. An exhaust flame trench ran through the center of the pad, and a wedge-shaped, heat-resistant flame deflector was moved into place below the first stage engines before lift-off. Since a tremendous amount of heat was produced by the powerful rocket at ignition, 45,000 gallons (170,325 liters) of water per minute were sprayed on the pad to cool the deflector, trench, and Mobile Launcher. Liquid oxygen, kerosene, and liquid hydrogen were stored nearby for fueling the vehicle prior to flight.

After unloading the Mobile Launcher with the Moon Rocket, the Crawler-Transporter traveled back down the Crawlerway to pick up the Mobile Servicing Structure, a 400 foot (122 m) tall steel-truss tower which partially enclosed the launch vehicle. Five work platforms gave technicians 360 degree access to both rocket and spacecraft until just 11 hours before lift-off when the Mobile Servicing Structure was carried back to its parking area alongside the Crawlerway.

Three and one-half hours before launch, the three astronauts rode an elevator up the umbilical tower of the Mobile Launcher to Service Arm 9 where they strolled across a 320 foot (97.5 m) high catwalk to their waiting spacecraft. All nine service arms swung out of the way at lift-off.

It may very well be that the first 3½ miles to the Moon would be the most difficult, but with innovation and hard work American engineers had created the infrastructure which would facilitate this initial portion of the journey. History's first Moonport was a reality.

The Mission Operations Control Room — better known as "Mission Control" — at the Manned Spacecraft Center in Houston, Texas. This was the facility from which ground controllers monitored the progress of each space flight. Huge maps displayed the locations of spacecraft, and a large television monitor at the front of the room displayed real-time images transmitted from space. Here controllers watch as Neil Armstrong prepares to step onto the lunar surface during the Apollo 11 mission. (NASA via Johnson Space Center)

8 The Great Train Wreck

After President Kennedy's speech, calling for a manned lunar landing by the end of the decade, America's aerospace industry mobilized its resources to meet the challenge. Across the country, private enterprise took on the responsibility of producing the hardware for Project Apollo. NASA's job was to coordinate that effort and train the astronauts and ground personnel who would carry out the mission. Training was done through a network of individual space centers — each focusing on a specific function.

The overall program was the responsibility of the Office of Manned Space Flight (OMSF) under the leadership of Dr. George E. Mueller at NASA Headquarters in Washington, DC. The network included:

- **The Manned Spacecraft Center** (renamed the Lyndon B. Johnson Space Center in 1973) at Houston, Texas, which was responsible for the development of the Apollo spacecraft, flight crew training, and flight control.

- **The George C. Marshall Space Flight Center** at Huntsville, Alabama (von Braun's organization, known as the Army Ballistic Missile Agency before it was transferred to NASA in 1960), which was responsible for developing the Saturn launch vehicles. The Marshall Space Flight Center also directed the **Michoud Assembly Facility** at New Orleans, Louisiana, where the first stages of the Saturn IB and Saturn V were produced, as well as the **Mississippi Test Facility** (renamed the John C. Stennis Space Center in 1988) in Bay Saint Louis, Mississippi, where the first and second stages of the Saturn V were test-fired and accepted for flight.
- **The John F. Kennedy Space Center** in Florida, which was responsible for launch operations.
- **The Goddard Space Flight Center** at Greenbelt, Maryland (near Washington, DC), which managed the Manned Space Flight Network that operated a global network of powerful antennas to track the Apollo spacecraft on its lunar mission and maintained communications throughout the flight.

Other centers contributing to the Apollo Project included NASA's **Ames Research Center** at Sunnyvale, California; the **Flight Research Center** at Edwards,

The Lunar Landing Training Vehicles (LLTVs) accurately reproduced the handling characteristics of the Lunar Module. Powered by a downward-pointing jet engine that simulated lunar gravity, the LLTVs were difficult to fly; Neil Armstrong was forced to eject from an LLTV the year before he set Eagle *down on the lunar surface. In this view, NASA test pilot Joe Walker maneuvers the LLTV at the Flight Research Center at Edwards, California. (NASA via Flight Research Center)*

From left to right, Astronauts Roger B. Chaffee, Edward H. White II and James A. McDivitt train in a Command Module simulator which accurately reproduced the spacecraft's interior. (NASA)

California; the **Langley Research Center** at Hampton, Virginia; the **Lewis Research Center** at Cleveland, Ohio; and the **Jet Propulsion Laboratory** at Pasadena, California (operated under NASA contract by the California Institute of Technology).

From lift-off to splashdown, each Apollo mission was controlled from the Mission Control Center at the Manned Spacecraft Center. Two mission control rooms were equipped with a variety of electronic subsystems, including communications equipment, displays, simulation and training facilities, and computers. The control rooms could support a manned flight and a simulated flight simultaneously.

Simulations were a vital part of training for both the astronauts and the mission controllers, preparing them for the unexpected, as well as the routine. In some cases, actual mission hardware was used for training. In other cases, mock-ups and simulators were utilized. Some of the more important simulators duplicated the cabins of the Apollo Command Module (CM) and Lunar Module (LM).

Astronaut John Young dubbed the CM simulator "the great train wreck" because from the outside it looked like a jumbled pile of boxes, wires, and unrelated contraptions. Inside, the simulator was an identical twin of an actual CM, with working dials, gauges, switches, controls, and panel lights. Simulated views of Earth, the Moon, and the stars were projected outside the windows. The astronauts could practice every step of their lunar voyage from lift-off to reentry, including rendezvous and docking. The LM simulator enabled Apollo flight crews to practice lunar landings without ever leaving the ground!

Ground crews also trained with these simulators from their consoles in the Mission Control Center. Computers created emergency situations that tested the skill of ground controllers and astronauts alike. Encountering multiple emergencies in practice would prepare the entire team for a genuine crisis during an actual mission.

Safely landing the LM on the Moon would be a unique challenge. The LM hovered somewhat like a helicopter, but handled rather differently. Its main thrust came from a powerful descent engine while steering was provided by small thrusters on each of the four Reaction Control System quads. To practice lunar landings, strange-looking machines called Lunar Landing Training Vehicles were built and flown. Flying on a column of jet engine exhaust, these "flying bedsteads" provided pilots with an authentic simulation of the LM's handling characteristics. But they were extremely tricky to fly. Apollo 11 astronaut, Neil Armstrong, ejected from one on May 6, 1968 (the year before the first Moon landing). Armstrong parachuted to safety while the Lunar Landing Training Vehicle crashed and burned.

Zero gravity could be simulated for up to 30 seconds, using Air Force transports following parabolic flight paths, while one-sixth gravity (to practice Moon walking) was reproduced through the use of slings and counterbalances.

Spaceships and spacesuits were tested in vacuum chambers which recreated the harsh environment of space. Mock-ups provided useful vehicles for practicing entering and exiting from the Apollo spacecraft.

Astronauts had to undergo survival training in case their returning CM touched down in some unplanned remote location.

The astronauts also studied geology. They took field trips to terrestrial sites that resembled their lunar targets and learned to identify rocks and minerals as well as geologic structure and terrain. They studied lunar science, too.

At the time the decision to go to the Moon was made, no one knew if human beings could survive in space for a week or more. The effects of weightlessness were largely unknown. Rendezvous and docking had never been attempted. The lives of the astronauts depended not only on astronauts, controllers, and support personnel practicing together until they melded into a team capable of flying the spacecraft almost in their sleep, but on anticipating and training for every conceivable mode of failure.

9 Mercury and Gemini: Earth Orbit

Project Mercury was a major steppingstone to Apollo. Its original objectives were to develop the basic hardware and technology needed to put a manned spacecraft into orbit around Earth, and to investigate man's ability to survive and perform in the space environment. These goals would be achieved using the simplest and most reliable means possible.

In part, Mercury had been conceived as an American response to the Russian Sputniks. It became an official NASA program on November 26, 1958, when the Space Task Group was formed at Langley Research Center in Virginia (later reorganized as the Manned Spacecraft Center at Houston, Texas). McDonnell Aircraft Corporation was given the Mercury production contract on February 5, 1959.

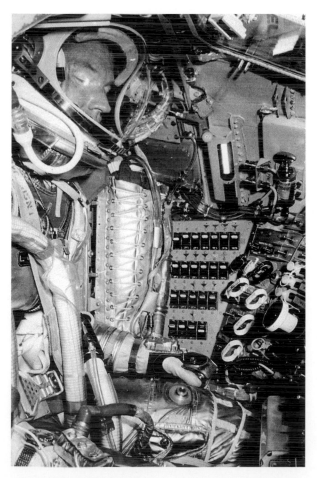

The cramped quarters of the Mercury spacecraft are evident in this view of Astronaut Gordon Cooper during a pre-flight test inside Faith 7. Mercury pilots could not float around their cramped cabins; merely climbing in and out of the "capsule" was an arduous task. The limited thrust of America's early rockets necessitated the use of small spacecraft. (NASA)

The initial sub-orbital missions used 78,000 pound (35,400 kg) thrust Redstone rockets, while orbital flights used 360,000 pound (163,300 kg) thrust Atlas missiles. The first unmanned test flights of the Atlas launch vehicle occurred in 1960, with several more being flown in 1961.

Although the Atlas D rocket was considerably more powerful than the Redstone, its thrust was still quite limited compared to the rockets the Soviets used to boost their orbital missions. Thus, the Mercury spacecraft weighed only 3,000 pounds (1,360 kg). It was slightly more than 9 feet (2.7 m) tall and just over six feet (1.8 m) wide at its base.

Two-stage Titan II missiles were used to boost the Gemini spacecraft into orbit. On August 21, 1965, Gemini 5 was launched from Cape Kennedy, sending Gordon Cooper and Charles "Pete" Conrad on an eight-day flight. (NASA)

Edward H. White II was the first American to "walk" in space. Gemini 4 Command Pilot James McDivitt took this memorable photograph of White floating just outside his window. The gold visor on White's helmet filtered the Sun's brilliant rays. (NASA)

The bell-shaped design was chosen so that the blunt end could act as a heat shield during reentry. At launch, a 17 foot (5.2 m), rocket-powered escape tower was mounted to the top of the spacecraft, providing a means to carry the capsule to safety in the event of a problem with the Atlas.

The decision to go to the Moon hardly seemed justified by Astronaut Alan B. Shepard's 15-minute, sub-orbital "hop" in a small, one-man Mercury spacecraft on May 15, 1961. After all, Commander Shepard had only achieved an altitude of 115 miles (184 km) and had experienced a mere five minutes of weightlessness. But Shepard's flight got Project Mercury off to a good start, despite the Soviet Union's dramatic success with the orbital flight of Yuri Gagarin the previous month.

On July 21, 1961, Astronaut Virgil I. "Gus" Grissom flew a second sub-orbital mission aboard *Liberty Bell 7*. (The Mercury astronauts named their own spacecraft, adding a "7" in recognition of their seven-man team.) Grissom's flight was a success, despite the fact that he nearly drowned when his capsule sank after splashdown in the Atlantic.

On February 20, 1962, the Mercury spacecraft was ready to meet its prime objectives. John H. Glenn, Jr. and his *Friendship 7* spacecraft were launched atop an Atlas for a three-orbit flight which lasted nearly five hours. It was a complete success and Astronaut Glenn became an overnight hero. Then on May 24, 1962, just three months after Glenn's historic flight, Astronaut M. Scott Carpenter repeated Glenn's mission in *Aurora 7*.

Gemini 6 astronauts, Wally Schirra and Tom Stafford, photographed Gemini 7 (with Frank Borman and Jim Lovell on board) during history's first space rendezvous on December 15, 1965. The two spacecraft came within six feet of each other and proved the feasibility of Lunar Orbit Rendezvous for future Apollo missions. The Gemini 7 astronauts spent a record-breaking 14 days in space. (NASA)

Unfortunately for the United States, the Soviet cosmonaut, Gherman Titov, had already spent a full day in space aboard Vostok-2, and in August 1962, *two* cosmonauts, Andrian Nikolayev and Pavel Popovich, were launched just one day apart in Vostok-3 and Vostok-4. Nikolayev spent nearly four days in space, Popovich nearly three.

On October 3, 1962, Astronaut Walter M. Schirra Jr. orbited Earth six times in his *Sigma 7* spacecraft and Astronaut L. Gordon Cooper brought Project Mercury to a very successful end with his 22 orbit, 34½ hour mission in *Faith 7* on May 15-16, 1963. One month later, the Soviet Union responded to America's success with another dual flight. Aboard Vostok-5, Cosmonaut Valeri Bykovsky was in orbit just short of five days, while Valentina Tereshkova, aboard Vostok-6, was the first woman in space. Her three-day mission exceeded the total time spent in space by all six Mercury astronauts.

On October 12, 1964, using a new third stage on the Vostok rocket, and a modified Vostok spacecraft, called Voskhod (Russian for Sunrise), Sergei Korolev struck again with the launch of the world's first three-man spaceship. (It would be four years before the United States could perform such a feat.) Cosmonauts Vladimir Komarov, Konstantin Feoktistov (a spacecraft designer), and Boris Yegorov (a physician), spent a full day in space. Their flight was more of a risky stunt than a technical achievement. However, on

March 18, 1965, Korolev scored a genuine space spectacular with the launch of Voskhod-2. On board were two cosmonauts, Pavel Belyayev and Alexei Leonov. Exiting Voskhod-2 through an inflatable airlock, Leonov became the first person to "walk" in space. This was a significant development in the Race to the Moon as it demonstrated that humans could work outside of their spacecraft in the vacuum of space.

Four years after Gagarin's single orbit and Shepard's 15-minute space hop, it was clear that humans could adapt to weightlessness and survive the stresses of launch and reentry. Project Mercury had given the United States the basic technology required for manned spaceflight, including a tracking network to control and monitor each mission, as well as the management skills to move forward with Apollo.

To bridge the gap between Mercury and Apollo, NASA launched Project Gemini on December 7, 1961. Gemini was a larger, improved two-man version of the basic Mercury capsule. Gemini's primary goals were to gain experience with extended periods of weightlessness, spacewalking, orbital rendezvous and docking, and precision reentry, all of which were prerequisites for Apollo.

The Agena target vehicle as seen by Gemini 8 astronauts Neil Armstrong and David Scott just before the first space docking on March 16, 1966. Shortly afterwards, a short circuit resulted in a thruster on Gemini 8 firing out of control, sending the combined Gemini-Agena into a wild spin. Armstrong quickly undocked, shut down his main thrusters, and activated a back-up system which necessitated an emergency return to Earth. (NASA)

The responsibility for transforming Mercury into Gemini fell upon the able shoulders of James A. Chamberlin, who had come to NASA from Canada, where he had worked on the Avro Canada CF-105 Arrow, which before its cancellation in February, 1959, had been regarded as the world's most advanced supersonic aircraft. Its cancellation resulted in the demise of Avro Canada, and NASA was quick to hire two dozen of the company's top engineers, including Chamberlin, who had been the Chief of Technical Design. While many people are aware of the major contributions of the German rocket scientists who had emigrated to America at the end of World War II, few recognize that the Canadians who joined the Space Task Group in 1959 also had significant influence on the Mercury, Gemini, and Apollo programs. Canada's loss proved to be America's gain.

Building on Mercury's proven technology, Chamberlin introduced important innovations to Gemini. Drawing on his experience with aircraft engineering, he introduced a modular approach, making individual components and systems more accessible and easier to maintain. This increased the reliability of the spacecraft and reduced the time needed to test and prepare it for flight. Accordingly, much of Gemini's equipment was housed outside the Reentry Module (astronauts' cabin) in a two-piece Adapter Module that would be jettisoned at the end of each mission.

The rear Equipment Section held propellant tanks, oxygen supplies, fuel cells (which replaced batteries for generating electrical power), a liquid-coolant radiator to dissipate spacecraft internal heat, and eight attitude control thrusters. The Retrograde Section housed the retrorockets used during reentry and additional maneuvering thrusters.

At its base, the Adapter Module was 10 feet (3.0 m) in diameter (to fit the Titan II launch vehicle) and 7.5 feet (2.3 m) at the top (to fit the Reentry Module's rear heatshield). The complete spacecraft stood 19 feet (5.8 m) tall, weighed 8,300 pounds (3,765 kg), and required the new 430,000 pound (195,000 kg) thrust, two-stage Titan II missile to reach Earth orbit. As with Mercury, McDonnell Aircraft was the prime contractor.

Starting with the three-orbit mission of Gemini 3 (the first two Gemini test flights were unmanned) on March 23, 1965, the program achieved all of its major goals over a period of twenty months. Veteran astronaut Gus Grissom and rookie John Young (the first of the new astronauts NASA was recruiting for Apollo) made history by altering their orbit in Gemini 3. On June 3, 1965, Gemini 4 lifted-off with James McDivitt and Edward White on a four-day mission, during which White became the first American to walk in space, and the first person to use a hand-held maneuvering unit to control his movements while in raw space.

In August, 1965, Gordon Cooper and Charles "Pete" Conrad spent eight days in space aboard Gemini 5, setting a new endurance record, and proving that men could live in space long enough to fly to the Moon and back (Apollo would take a minimum of eight days to make its round-trip lunar voyage).

To practice rendezvous and docking, NASA planned to put unmanned Agena-D rockets into orbit, launching them atop Atlas missiles. Radar would help the Gemini astronauts find their targets, and Gemini's nose would slip into a cylindrical collar mounted on the nose of the Agena. Special locks would hold the vehicles firmly together, and the Agena's powerful rocket engine could be re-ignited to send the docked spacecraft into a higher orbit.

When the Agena target vehicle failed to reach orbit on October 25, 1965, Gemini astronauts Walter Schirra and Thomas Stafford were handed a new assignment for Gemini 6; the first space rendezvous. On December 15, 1965, Schirra and Stafford met Frank Borman and James Lovell, who were already in space aboard Gemini 7. Gemini 6 came home the next day, but Borman and Lovell (who had been launched on December 4) stayed in orbit for a record-breaking two weeks! It was now proved beyond any doubt that men could live and work in space for extended periods.

History was made again on March 16, 1966, when Neil Armstrong and David Scott achieved the first space docking with an Agena. Success was threatened by near-tragedy when one of the thrusters on the Gemini 8 spacecraft wouldn't shut off, sending the two vehicles into a wild spin. Armstrong coolly broke free from the Agena and reestablished control of his spacecraft. The emergency forced the astronauts to make an unplanned splashdown in the Pacific, but disaster had been avoided and valuable lessons had been learned.

Between June and November, 1966, four more Gemini flights were made in quick succession: Thomas Stafford and Eugene Cernan in Gemini 9, John Young and Michael Collins in Gemini 10, Charles Conrad and Richard Gordon in Gemini 11, and James Lovell and Edwin Aldrin in Gemini 12. These missions perfected the complex techniques required for orbital rendezvous and docking, controlled reentry, and spacewalking.

By the end of 1966, NASA astronauts and mission planners had acquired a wealth of experience in manned spaceflight. The Moon was getting closer all the time.

10 Ranger: Live from the Moon!

At its peak during the mid-sixties, an estimated 430,000 Americans were directly involved with putting a man on the Moon. These scientists, engineers, technicians, and specialists were from a host of fields and disciplines. Many of them were involved with building the hardware that would carry the Apollo astronauts to the Moon. But before the Lunar Module (LM) could make its first landing, there was a great deal more to learn about the Moon itself. Therefore, some of those 430,000 men and women were busy designing, building, and testing a series of unmanned lunar scouts—robotic spacecraft. The scouts' names defined their missions—Ranger, Surveyor, and Lunar Orbiter.

Ranger 9 was launched atop an Atlas-Agena B rocket on March 21, 1965. (NASA)

The dictionary defines a "ranger" as someone who patrols or reconnoiters a specific area or region. Project Ranger was designed, in part, to reconnoiter specific regions of the Moon's surface along a narrow strip, centered on the lunar equator. This "Apollo Zone" was targeted because it was the most accessible area for a manned landing, and because it was thought to be relatively smooth. Yet the question remained—just how smooth was it? And could a spacecraft land there safely? Out of nine Rangers, the last three provided some partial answers. But getting them to the Moon in the first place proved to be a frustrating process.

Project Ranger was born in December, 1959, two months after the Soviet Union had taken the first pictures of the Moon's far side with Luna-3. The Jet Propulsion Laboratory (JPL) in Pasadena, California was chosen to manage Ranger, which became the first in a long series of headline-grabbing lunar and planetary missions carried out by JPL over the years (others included Mariner, Surveyor, Voyager, and Galileo). Among its many objectives, Ranger was supposed to transmit the first close-up views of the Moon's surface, showing details as small as 10 feet (3.0 m) across.

Unfortunately, Rangers 1 and 2 were failures. Rangers 3, 4, and 5 carried sophisticated scientific experiments, in addition to cameras. While all Ranger spacecraft were designed to crash into the Moon, Rangers 3, 4, and 5 were equipped with a small, spherical instrument package encased in a 2 foot (0.6 m) diameter balsa wood ball. Some 70,000 feet (21,335 m) above the lunar surface, the ball would be ejected from the carrier spacecraft. A retrorocket attached to the bottom of the ball would fire, slowing it enough so that the package inside could survive a hard landing at just under 150 mph (240 kph). A few minutes later, a sensitive lunar seismometer would start recording Moonquakes and relaying other data about the Moon. Rangers 3, 4, and 5 fared no better than Rangers 1 and 2, although Ranger 4 crashed into the Moon's far side, making it the first American spacecraft to reach the lunar surface on April 26, 1962. The failures of these early lunar spacecraft stemmed from two main causes. Because of the payload limitations of the Atlas-Agena B launch vehicle, they were built without back-up systems to save weight. They were also put through a sterilization process to protect the Moon from possible Earth contaminants, a process which took its toll on Ranger's sensitive electronic components.

Ranger 6 was launched on January 30, 1964. It carried only television cameras, designed to transmit high-resolution pictures of the Moon before the spacecraft plunged into the Moon's surface at 6,000 mph (9,600 kph). Having a single objective did make the mission less complicated. Nevertheless, Ranger 6 was another failure. Although it flew straight to its target in the Sea of Tranquillity, an electrical short that occurred at launch burned out portions of the camera system.

During the following six months, engineers took a number of measures to prevent a recurrence of the Ranger 6 problem. Finally, on July 28, 1964, Ranger 7 left Cape Kennedy. The Agena put the spacecraft into a parking orbit around Earth, then sent it on

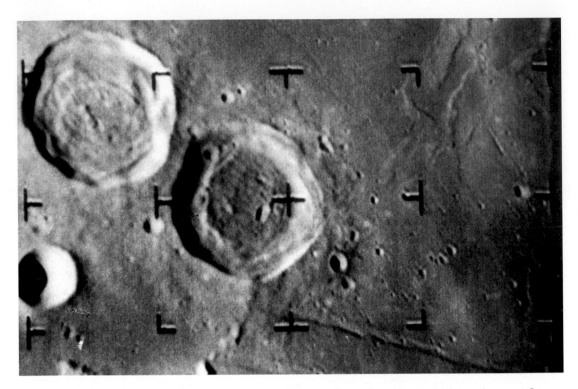

Taken just 151 miles (243 km) above the Moon by Ranger 8, this photograph shows the twin, flat-bottomed craters Ritter and Sabine as well as the parallel Hypatia Rilles. The resolution is about ten times better than the best previous Earth-based views. The rilles resembled terrestrial graben — narrow blocks of material dropped downward (relative to the rocks on either side) between two parallel faults. Some scientists thought Ritter and Sabine looked like volcanic calderas, while others suggested they were caused by the simultaneous impact of two asteroids. Each is approximately 18 miles (30 km) in diameter. (NASA)

a precision course to the Moon. On July 31, 1964, Ranger 7 started transmitting the first of 4,308 clear, close-up pictures of the Moon — the final images being 1,000 times more detailed than any recorded by Earthbound telescopes. It was a stunning success, and in its honor, the unnamed sea in which Ranger 7 crashed was named Mare Cognitum (the Known Sea).

Scientists who expected to see deep craters, jagged cliffs, and huge crevices were surprised by the gently rolling terrain revealed by Ranger 7. Although an astounding number of craters were seen in the photographs, overall the Moon appeared remarkably smooth. In one of the final frames, a few angular, boulder-sized rocks could be seen at the bottom of one crater. NASA planners could assume that any surface which could support such boulders could also support the weight of a LM.

The craters seen in the Ranger photographs ranged in size from hundreds of miles down to the limit of visibility, which seemed to confirm the hypothesis of Dr. Eugene M. Shoemaker, Chief of the Astrogeology Branch of the U.S. Geological Survey, that primary craters were formed by large meteorites striking the Moon at high velocity, while secondary craters were formed by chunks of rock thrown out of the primaries. The photographs seemed to confirm that the mysterious lunar rays were actually chains of secondaries created by debris thrown up from the lunar surface by the impact which had created primary craters such as Tycho and Copernicus.

Ranger 8 returned another 7,137 lunar pictures before crashing into the Sea of Tranquillity (not far from the eventual Apollo 11 landing site) on February 20, 1965. Many of the images showed the rugged lunar highlands. The twin flat-bottomed craters, Ritter and Sabine, looked to some geologists like giant calderas, re-igniting the controversy over volcanic versus impact crater formation.

Ranger 9 concluded the series on March 24, 1965, when it impacted the Moon in the crater Alphonsus, which has features that clearly seem to be of volcanic origin. Many of Ranger 9's 5,814 pictures were broadcast on Earth in real time, with the words, "Live from the Moon," appearing for the first time on home television sets. A new era in lunar exploration had begun.

11 Lunar Orbiter: Mapping the Moon

The results of Project Ranger were not entirely clear in 1965. Although the scientists studying Ranger images pretty much agreed that the three regions targeted by Rangers 7, 8, and 9 were smooth enough for a lunar landing and that the surface was fairly solid, they still argued about the origins of specific features and how to interpret many of the details seen in the photographs.

The astrogeologists wanted to observe larger areas in equal or even greater detail and NASA wanted them to map those regions of the Apollo Zone that offered the safest landing sites. The best way to achieve these goals would be to place a spacecraft equipped with both wide-angle and telephoto camera lenses into orbit around the Moon to photograph specific targets in pre-selected areas. The Lunar Orbiter project, which was managed by the Langley Research Center at Hampton, Virgina, was assigned this task.

Lunar Orbiter started out as an offshoot of the Surveyor program, which had been approved in 1960. This was before President Kennedy had anointed Apollo with the task of a manned lunar landing, so Surveyor had the starring role in the scientific investigation of the Moon. Two versions of Surveyor using common hardware were envisioned; an orbiter and a lander. The first would survey broad areas from above, while the second would perform detailed studies on the surface below.

After Apollo, Surveyor moved into a supporting role, its new assignment being to assess the lunar environment from an engineering, rather than a scientific point of view.

The focus of Surveyor was now on the lander, and plans for the orbiter version were shelved. Yet an orbital reconnaissance spacecraft was still needed for the mapping requirement. So Lunar Orbiter emerged as an entirely new project in 1963. Langley prepared the design specifications, and the Boeing Company won the prime contract to produce the spacecraft. The proven Atlas-Agena D was chosen as the launch vehicle.

In order to obtain the most detailed pictures possible—showing objects as small as 3 feet (0.9 m) across—NASA engineers decided to use high-definition, 70 mm photographic film rather than television imaging. A dual-lens camera offered medium-resolution photographs (80 mm lens) of broad regions, and high-resolution photographs (610 mm lens) of specific targets. Due to the slow speed of the film and the spacecraft's high speed over the lunar surface, a device was provided to eliminate image blurring. A special V/H Sensor measured Lunar Orbiter's velocity and height, then signaled an image motion compensation mechanism which advanced the film slightly (just the right amount) during each exposure.

Eastman Kodak was responsible for Lunar Orbiter's camera, including the on-board Bimat film processor. Housed inside a pressurized, temperature-controlled container, the processor automatically developed the film and sent it to an optical scanner where the negatives were converted into electronic signals for transmission to Earth. It was a

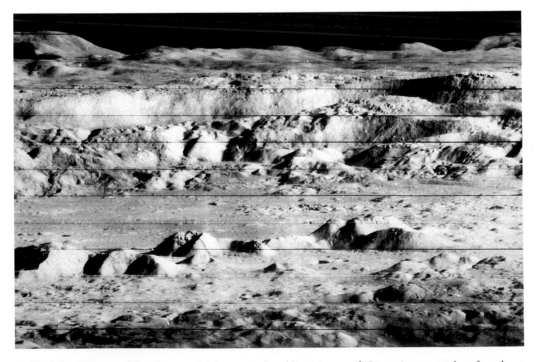

Dubbed the "Picture of the Century," this spectacular oblique image of Copernicus was taken from lunar orbit by Lunar Orbiter 2. The crater's central peaks can be seen just below the center of the picture, while the Carpathian Mountains are visible in the distance, outlined against the black lunar sky. (NASA)

In this photo from Lunar Orbiter 5, the crater Tycho, 53 miles (85 km) in diameter, is the main feature. A relatively young crater (estimated to be 110 million years old), Tycho still reveals much of the "ejecta" (surface material) that was thrown out by the impact that formed it. A rugged area some 18 miles (29 km) beyond Tycho's north rim (near the top of this picture) was chosen as the landing site for the Surveyor 7 spacecraft. (NASA)

complex system, but it produced the high-resolution reconnaissance images that were necessary for detailed lunar mapping. The spacecraft also carried instruments to measure radiation and to detect micrometeroids in lunar orbit—in order to determine what hazards, if any, could be expected by Apollo astronauts in the vicinity of the Moon. Two antennas and four paddle-shaped solar panels (for electrical power) were mounted around the exterior of Lunar Orbiter, and a velocity control engine provided the thrust for orbital insertion and adjustment (the same type of engine used in the reaction control systems of the Apollo Service and Lunar Modules). During a typical mission, the standard elliptical orbit took the spacecraft from 1,150 miles (1,850 km) at its high point down to just 28 miles (46 km) above the surface.

Lunar Orbiter 1 left Cape Kennedy on August 10, 1966, and was circling the Moon four days later. Its first pictures were taken on August 18, and it soon became obvious that there was a problem with the V/H Sensor, blurring most of the high-resolution images. But Lunar Orbiter 1 returned some 207 medium-resolution photographs (unaffected by the faulty V/H Sensor), showing Apollo Zone targets, as well as detailed images of the Moon's far side. One picture, in particular caught the world's imagination. It was the first photograph of the home planet taken from a spacecraft in the vicinity of the

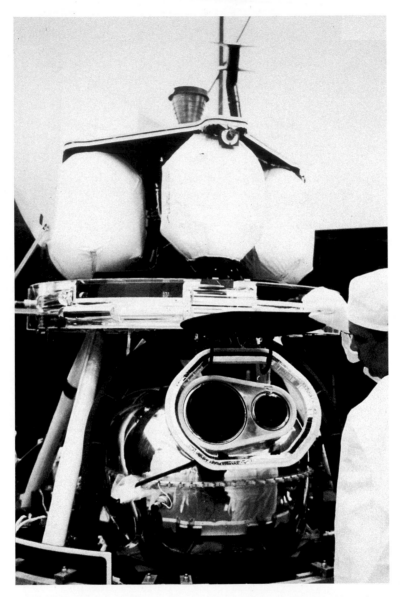

A technician checks the dual lenses of Lunar Orbiter 5 prior to launch. He is holding the camera's thermal door, which acted as a lens cap during the mission. Propellant tanks and the velocity control engine nozzle can be seen at the top of the spacecraft. (NASA)

Moon. Showing a cloud-covered, crescent Earth hanging over a barren, lunar landscape, it offered a preview of more dramatic scenes to come.

On October 29, a command was sent to Lunar Orbiter 1 that sent it crashing into the Moon, thereby avoiding any communications interference with Lunar Orbiter 2 which was launched on November 6, 1966. Once again, the Apollo Zone was the main target of the orbiting probe's camera lenses. And this time, there were no problems with the image motion compensation mechanism. The spacecraft returned incredibly sharp photographs of the lunar surface, including a truly breathtaking view of the crater, Copernicus, that the press hailed as the "Picture of the Century." It showed the 2,000 foot (610 m) central peaks and terraced walls of Copernicus, with the Carpathian Mountains in the distance — all set against a jet black lunar sky. For the first time, the Moon was seen from a familiar perspective. It was still an alien landscape, to be sure, but not all *that* alien.

Lunar Orbiter 3 left Earth on February 5, 1967. Like its immediate predecessor, it returned high-quality images of the Apollo Zone, successfully concluding Lunar Orbiter's mapping function. Some of the potential Apollo landing sites were shown to be too rough, while others seemed quite acceptable for both the unmanned Surveyor series and Apollo. Indeed, Lunar Orbiter 3 photographed the Surveyor 1 spacecraft resting on the lunar surface (it had landed the previous June, two months before the first Lunar Orbiter). It also obtained extraordinary close-up views of meandering tracks left by huge boulders rolling down crater walls, and other images of scientific interest to lunar geologists.

Having achieved the program's main objectives, the last two Lunar Orbiters were dedicated purely to science. The goal of Lunar Orbiter 4, launched on May 4, 1967, was to photograph the entire near side of the Moon. Included among its 163 images were some spectacular wide-angle pictures of the Orientale Basin on the Moon's far side. This huge, multi-ringed, impact basin, looking much like a giant bull's eye, shed new light on the geologic history of the Moon. Orientale appeared to be the youngest of the large, circular lunar basins, and was only partly filled with dark volcanic material (unlike the fully-covered Mare Imbrium or Mare Crisium on the near side). Its four, well-preserved mountain rings would eventually help scientists explain many of the strange features seen on the lunar surface.

Lunar Orbiter 5 was sent to the Moon on August 1, 1967. It returned 212 photographs. The combined images from all five missions covered 99 percent of the Moon's surface. Several penetrating views of craters, such as Tycho and Aristarchus, proved to be scientifically significant as well as aesthetically pleasing. The Lunar Orbiter program concluded on a high note.

One other important contribution came from the Lunar Orbiter program. During the first mission, small deviations were observed in the spacecraft's orbit around the Moon. The cause was later traced to mass concentrations — termed mascons — of material buried beneath the maria, creating minor changes in the lunar gravity field. Mascons would have to be taken into account when planning orbital maneuvers for the Apollo flights.

Thanks to Lunar Orbiter, the first men on the Moon would have road maps to guide their way.

12 Surveyor

Sergei Korolev was the Soviet Union's anonymous Chief Designer of Rockets and Space Systems. His identity was revealed to the world—finally—when he died prematurely at age 59 on January 14, 1966. In death, he finally received the public recognition that he had deserved in life. Korolev's passing was a major blow to his country's plan to land a man on the Moon. Yet, less than three weeks later, the Soviet Union captured the headlines once again with another space first. This time the official announcement was just two lines long:

> **Moscow, February 3.** *'Luna-9' has accomplished a soft landing on the Moon's surface. The Soviet station 'Luna-9,' which has soft-landed on the Moon, has a reliable communication link with Earth.*

This brief report underestimated the true significance of the spacecraft's historic achievement; at last, mankind could view the Moon from the lunar surface itself.

Although described as "soft," Luna 9's landing was anything but a smooth, gentle touchdown. Had a cosmonaut been on board, he would not have survived the impact. Just before the main spacecraft crashed into the surface, the landing capsule was ejected—much like the balsa wood balls that had been planned for Rangers 3, 4, and 5. Luna-9's probe was protected by two shock-absorbing balloons which enclosed the egg-shaped capsule. The balloons were discarded after landing, and four metal petals opened up, exposing a rotating camera which produced a series of crude panoramas. Only a few small rocks and shallow craters were discernible. Nevertheless, Luna-9 proved that a spacecraft could land on the Moon and that the Moon's surface was indeed relatively smooth. It also stole the thunder from America's far more sophisticated Surveyor series. But not for long.

The first of seven Surveyor robot spacecraft was launched from the Cape atop an Atlas-Centaur rocket on May 30, 1966. The Centaur second stage was powered by two RL-10 engines (the first engines to successfully operate in space using high-energy liquid hydrogen as a fuel). Although plagued by failures and technical snags during its protracted development, Centaur went on to become one of NASA's most important upper stages. Atlas-Centaur Number 10 put Surveyor Number 1 on a precise heading toward the Moon.

During the cruise portion of Surveyor's mission, and while on the lunar surface, electrical power was provided by a large solar panel mounted on a mast at the top of the triangular-shaped spacecraft. Also attached to the mast was a High-Gain antenna for transmitting high-resolution television pictures. Communications with the vehicle were maintained via two transmitters, two receivers, and two omnidirectional antennas affixed to the main frame. In addition, Surveyor 1 carried two thermally-controlled compartments for electronic equipment, as well as an auxiliary storage battery, two television

Surveyor 1 during check-out prior to launch. The craft's legs and antennas are folded, and its remotely controlled television camera can be seen at left center. The fixed downward-looking camera (not used during the mission) can be seen at left beneath the folded leg and footpad. (NASA)

cameras, assorted fuel tanks for the attitude control jets and vernier rocket engines, two radar antennas, three landing gear legs (with shock absorbers), and a solid fuel main retrorocket engine. Three rate gyros, a Sun sensor, and a star sensor kept Surveyor on course. Hughes Aircraft Company produced the spacecraft.

The accuracy of Surveyor 1's trajectory was such that only one minor mid-course correction was required, about 16 hours after launch. The target, near the crater Flamsteed, was a smooth area encircled by hills and low mountains that marked a ghost crater — an ancient depression flooded by lava flows in the distant past, leaving it barely discernible today. Half an hour before its scheduled landing, the spacecraft aligned its main retrorocket along the approach path. Some 47 miles (75 km) above the surface, traveling at 5,840 mph (9,400 kph), the retrorocket was ignited for a 38 second burn. During that time, Surveyor slowed down to 267 mph (435 kph) at which point the spent retrorocket motor case was ejected. The three liquid fuel vernier engines which had stabilized the

Surveyor 1 was launched from Complex 36A at Cape Kennedy on May 31, 1966. An Atlas-Centaur rocket put the spacecraft on course for a lunar touchdown three days later. (NASA)

craft during main retro fire continued to slow Surveyor's descent. Using altitude data supplied by the radar system, the verniers shut down about 10 feet (3.0 m) above the Moon. From that height, the spacecraft free-fell to the surface, touching down at 7.5 mph (12 kph).

Surveyor 1 reached the Moon at 2:17 A.M. on June 2, 1966. Although its shock absorbers caused the spacecraft to rebound a few inches off the ground, a pair of Apollo astronauts could have survived the touchdown, giving Surveyor 1 the distinction of making the first, true soft lunar landing. Half an hour after landing, Surveyor began shooting an initial group of 14 low-resolution (200-scan-line) pictures, transmitted via the omnidirectional antennas. The photos clearly showed that the spacecraft had touched down on a smooth surface and was undamaged.

After the High-Gain antenna was aligned with Earth, the first high-resolution (600-scan-line) images were beamed to the Space Flight Operations Facility at the Jet Propulsion Laboratory in Pasadena. Over the next six weeks, Surveyor 1 took more than 11,200 detailed pictures of the lunar surface, some of them more than a million times finer than the best photographs taken through Earth-based telescopes. They revealed a gently rolling landscape, studded with shallow craters and littered with rocks and fragments of various shapes, sizes, and textures. Rounded mountain crestlines hugged the horizon. Many of the pictures were self-portraits, including close-ups of the footpads, which confirmed that the powdery soil was similar to wet beach sand, or Portland cement, as had been predicted by astrogeologist Gene Shoemaker, one of the principal scientists NASA had assigned to Project Surveyor.

Surveyor 1 was equipped with two television cameras: a fixed, downward looking approach unit that was carried but not used, and a survey camera, mounted at eye-level on the spacecraft frame. The survey camera pointed upward, but a mirror assembly, housed at the top, could be tilted and rotated to allow the camera to look up, down, and around. The heart of the survey camera was its f/4.0 zoom lens which focused from

One of the first high-resolution images from the Surveyor 1 spacecraft showed a close-up of one of its footpads and the fine texture of the lunar soil. The pad made only a slight impression in the powdery soil, demonstrating its ability to support the weight of a lander. A calibration chart (upper right) was mounted on one of the spacecraft's legs for use with the craft's television camera. (NASA)

four feet (1.2 m) to infinity. The focal length was adjustable from 25 mm, for wide-angle views, to 100 mm for close-ups. The focal-plane shutter could be held open for time exposures, and a filter wheel enabled Surveyor 1 to take the first color pictures of the lunar surface, which turned out to be gray. Bell & Howell built the zoom lens.

Although Surveyor 1 was not designed to survive the cold, 14-day lunar night, after hibernating, it awoke on July 6, 1966, and sent another 618 images back to Earth. During the long night, one of the 36 mirrored glass thermal radiators covering the top of electronic Compartment A shattered, the result of lunar temperature extremes. Otherwise, Surveyor 1 came through in fine shape. Despite a faltering battery, the spacecraft continued to "phone home" off and on through January, 1967. This achievement was all the more remarkable considering the fact that, given the string of failures with Ranger, few people had expected the first Surveyor to so thoroughly succeed.

Surveyor 2, launched on September 20, 1966 toward Sinus Medii (Central Bay) in the center of the Moon was a complete failure. It crashed into the Moon's surface, southeast of Copernicus.

Surveyor 3 made history twice. The first time came with its launch on April 17, 1967. Unlike its predecessors, Surveyor 3 was put into a parking orbit around Earth before the Centaur engines were reignited, sending the vehicle on its way to the Moon. Once there, Surveyor 3 landed on the east wall of a 650 foot (200 m) diameter crater south of Copernicus and east of Flamsteed in the Ocean of Storms. The spacecraft

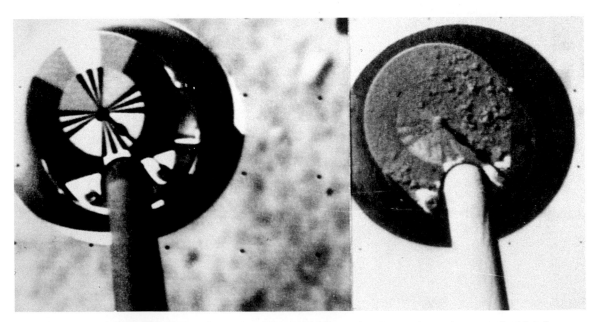

Surveyor 6 was the first spacecraft to take off from the Moon when its vernier engines were fired briefly, allowing a short "hop" across the lunar surface. As it did so, lunar soil beneath the engines was blasted upwards. In these before and after views, the soil was shown to be fine-grained and adherent, qualities that would cause Moonwalking astronauts to complain about their dirty spacesuits. (NASA)

actually touched down three times because its vernier engines failed to cut off automatically and the vehicle bounced twice before a command from Earth finally stopped the runaway rockets. Although its position inside and below the rim of a typical lunar crater prevented the television camera from viewing the terrain beyond the depression, one of Surveyor 3's main objectives was to take pictures of an area right in front of the spacecraft. This is where its new Soil Mechanics Surface Sampler pounded, scraped, and dug into the soil to test its key physical properties: bearing strength, texture, and structure.

Mounted beneath the survey camera (where the downward-looking camera had been positioned on Surveyors 1 and 2) the arm of the Soil Mechanics Surface Sampler could reach over 5 feet (1.5 m) away and dig a trench 18 inches (0.45 m) deep. It could also pivot through 112 degrees, enabling the small scoop at the end of the device to sample a fairly wide area. Both rocks and soil could be picked up, dropped, and moved about with this scoop. Surveyor 3 dug a total of 4 trenches and made 7 bearing tests and 13 penetration tests. The survey camera observed each of these experiments, including an attempt to crush a small rock which proved to be "rock solid!" (Some scientists thought the rocks might be clods of soil that would crumble apart when crushed.)

Although Surveyor 3 did not survive the lunar night, it did return 6,315 pictures during its two weeks of operation, and further demonstrated the ability of the lunar regolith (a geologic term Dr. Shoemaker used during this mission to describe the Moon's fragmented topsoil) to support the weight of an Apollo Lunar Module and Moonwalking astronauts. Thirty-one months later, Surveyor 3 would again make headlines when the Apollo 12 astronauts visited the robot lander and removed its scoop and survey camera. Although scientists now knew something about the physical properties of the Moon's regolith, they were still uncertain of its chemical composition. Another Surveyor was being readied to help unlock those secrets.

The Moon's Central Bay was the target once again when Surveyor 4 lifted off from Cape Kennedy on July 14, 1967. Unfortunately, all communications with the spacecraft were abruptly lost during the main retro fire.

Surveyor 5, launched on September 8, 1967, was aimed toward the Sea of Tranquillity, where it landed three days later. For a while, it looked like Surveyor 5 might be yet another failure. But a critical leak in a helium tank used to pressurize the vernier engine propellant tanks was overcome by some innovative, real-time changes in the flight plan. Once again, the spacecraft touched down in a crater. This time, the crater was only 30 feet (9.1 m) in diameter, and the camera could see just over the rim. The terrain was similar to that observed at the previous sites. Over 19,000 images were eventually transmitted to Earth.

As with Surveyor 3, one of Surveyor 5's main objectives was to sample the local regolith. In the case of Surveyor 5, the soil was analyzed chemically. Replacing Surveyor 3's arm and scoop was a small, gold-covered box, sitting on top of a white plate. This Alpha Scattering Experiment, designed under the leadership of Dr. Anthony Turkevich of the Fermi Institute at the University of Chicago, was lowered to the surface by a nylon line. An opening at the bottom permitted alpha particles from a radioactive source, Curium 242, to bombard the soil. Some of these particles were scattered back up into the box where six detectors measured their number and energy. Since different elements produce different results, an analysis of the alpha particles enabled scientists to determine several of the key constituents present in the soil at one mare location. This first, on-site chemical analysis of another celestial body was a historic achievement, although the results were not surprising. "The most abundant elements," Dr. Turkevich reported, "were oxygen, silicon, and aluminum, in decreasing order. The relative amounts of the elements resemble those in terrestrial basalts." In other words, the dark mare soil appeared to be of volcanic origin, as had been expected.

Surveyor 6, launched on November 7, 1967, broke the jinx associated with Sinus Medii and successfully landed in Central Bay on November 9. The spacecraft came to rest on level ground, although a mare wrinkle ridge was visible on the horizon about half a mile (0.8 km) away. The soil analysis at the Surveyor 6 mare site, made with another Alpha Scattering instrument, provided similar results to those obtained by Surveyor 5. Surveyor 6 returned some 30,000 images to Earth. It was a nearly flawless mission which also made history as the first spacecraft to lift-off from the Moon, taking a short hop across the surface, which enabled scientists to obtain stereo imagery of the surrounding terrain.

Having scouted four potential Apollo landing sites, and now satisfied that the Apollo Zone was safe for a manned lunar landing, NASA flight planners sent the last Surveyor on a purely scientific mission to the rugged lunar highlands surrounding the bright rayed crater, Tycho. Surveyor 7 left Earth on January 7, 1968, and landed 18 miles (29 km) north of Tycho's rim on January 10. While the maria represented relatively "young" terrain, the material blasted out from deep inside the Moon by the impact that created Tycho was thought to be quite old. The surrounding landscape was very different from previous landing sites with its hills and gullies, and many cracked, blocky angular rocks. The spacecraft carried both a Surface Sampler and an Alpha Particle device to analyze the regolith in this comparatively rough region of the Moon.

On the whole, the highland soil demonstrated the same physical properties as the mare soil at the Surveyor 3 site, but the chemical analysis showed a lower content in the iron group of elements (titanium, vanadium, chromium, manganese, iron, cobalt, nickel, and copper). Dr. Shoemaker would later identify this coarse-grained material as "anorthositic gabbro" composed of minerals (mostly plagioclase) formed deep beneath the ancient lunar crust.

Five highly successful Surveyor flights allowed us to "touch" the lunar surface via robotic instruments.

Now it was man's turn to fly to the Moon.

13 The Fire and the Phoenix

By the end of 1966, Surveyor had landed on the Moon, Lunar Orbiter was mapping its surface, and Gemini had demonstrated the techniques required to send spacemen to the Moon. NASA planned to launch the first manned Apollo Command/Service Module (CSM) combination, the first unmanned Lunar Module (LM), and the first unmanned Saturn V Moon Rocket by the end of 1967.

Until the Saturn V was "man-rated"—deemed safe for astronauts to ride—the smaller Saturn IB would be used for Apollo's Earth orbital test flights. NASA hoped to launch the second Apollo crew atop one Saturn IB, followed by the launch of an unmanned LM atop another IB. The CSM would rendezvous and dock with the LM. Then two of the astronauts would transfer to the LM and test its various systems in flight. It was an ambitious schedule, but a necessary one if a manned lunar landing was to be achieved by the end of the decade.

"Apollo 204" was NASA's designation for the first manned mission, subsequently called Apollo 1. The astronauts assigned to the flight were: Virgil "Gus" Grissom, command pilot; Edward White, senior pilot; and rookie Roger Chaffee, pilot. They were to fly a Block I spacecraft—an early model not designed for docking, intended only for

The crew of Apollo 1 (left to right): Edward H. White II, Senior Pilot; Virgil I. Grissom, Command Pilot; and Roger B. Chaffee, Pilot. (NASA)

Earth orbital test missions. Since it did not represent the definitive Block II configuration, it was the only ship of its type scheduled to carry a crew. But Spacecraft 012 (the production number) had a problem-plagued history.

Many of Spacecraft 012's troubles centered around its Environmental Control System which regulated the atmosphere, pressure, and temperature within the Command

A flash fire during a countdown demonstration test on January 27, 1967, took the lives of Apollo 1 astronauts Gus Grissom, Ed White, and Roger Chaffee. Here the scarred exterior of the spacecraft is seen shortly after the accident. (NASA)

Module (CM). The Environmental Control System had experienced several malfunctions in the weeks and months following the vehicle's arrival at the Kennedy Space Center on August 26, 1966. Managers and astronauts alike, expressed their dissatisfaction. Finally, by January 27, 1967, everything seemed ready for the Plugs Out test, a full dress rehearsal for the planned February lift-off.

The crew entered Spacecraft 012 early that afternoon. They were sealed inside by three hatches: an inner and an outer spacecraft hatch, and a hatch that was part of the Boost Protective Cover. None could be opened quickly or easily in an emergency. Once the cabin was sealed, it was filled with pure oxygen at 16 psi (1125 grams/sq cm). Locked inside, the astronauts were surrounded by Velcro® cloth (used to keep gear secured in zero gravity), plastic, rubber, and paper. Incredibly, no one had recognized how lethal this combination would be if a fire were to break out. A spark would turn the cabin into a bomb, which is exactly what happened at 6:31 P.M. when a crewman reported, "We've got a fire in the cockpit!"

Although they struggled bravely to save themselves, all three astronauts died of asphyxiation within seconds. At the same time, the pressure inside the spacecraft rose past the design limits and the cabin ruptured, pouring smoke and flames into the surrounding service structure, impeding the attempts of pad personnel to rescue the crew.

But nothing could have saved Grissom, White, and Chaffee. Their spaceship was fatally flawed. Before Apollo could be flown to the Moon or even into Earth orbit, engineers would have to make sure that it was safe, even on the ground.

The Moon Race claimed yet another life just three months later. Veteran Soviet cosmonaut Vladimir Komarov rode the new Soyuz (Russian for Union) spaceship into orbit on April 23, 1967. Like Apollo, Soyuz was designed to go to the Moon. And like Apollo, lunar orbit rendezvous was the mode selected to get there. But Komarov ran into trouble: Soyuz-1 was having control problems. A decision was made to bring him back to Earth the next day. Although Komarov re-entered the atmosphere successfully, his spinning spacecraft fouled the lines of the main parachute. Soyuz-1 plummeted to the ground, killing the highly-respected cosmonaut.

In the United States, an Apollo 204 Review Board had been created to investigate the cause of the January fire, and to recommend corrective actions based on its findings. The Board was headed by Dr. Floyd L. Thompson, Director of NASA's Langley Research Center, and included astronaut, Frank Borman. It could not identify a specific source for the fire, though an electrical arc in or near the Environmental Control Unit was considered to be the most likely cause. Contributing to the rapid spread of the fire was the pure oxygen environment and the abundance of flammable items within the cabin. New safety procedures were adopted. Fireproof materials were used throughout the redesigned, Block II CM and in the astronauts' spacesuits. A new, quick-opening hatch replaced the cumbersome three-piece unit.

On October 11, 1968, Apollo 7 rose phoenix-like from its launch pad, carrying astronauts Wally Schirra, Donn Eisele, and Walter Cunningham. (There was no Apollo 2 or 3; Apollos 4 to 6 were unmanned test flights.) The first manned Apollo flight lasted nearly 11 days. After 163 orbits around Earth, the astronauts splashed down in the Atlantic Ocean, and the "new" Apollo Moonship had passed a critical test.

The Apollo program got back on track with the launch of Apollo 7 atop a Saturn IB rocket on October 11, 1968. This was the first manned test of the Apollo spacecraft in Earth orbit. (NASA)

The Apollo 7 crew relaxes aboard the aircraft carrier USS Essex *at the end of their 11-day mission. Left to right are: Commander Walter M. Schirra, Jr.; Lunar Module Pilot R. Walter Cunningham; and Command Module Pilot Donn F. Eisele. Despite Cunningham's title, Apollo 7 did not carry a Lunar Module.* (NASA)

While the mechanics of the spacecraft worked perfectly, the men did not. Schirra and his crewmates contracted severe head colds which made them testy and short-tempered. But like good showmen everywhere, they put aside their misery long enough to turn in several highly-entertaining television performances from "the lovely Apollo Room, high atop everything!" Their good humor during the broadcasts served to restore the public's confidence in the fire-scarred Apollo program. And the performance of their spacecraft put America back on course toward a Moon landing by the end of the decade.

The Soviets were recovering their momentum as well. Four days after Apollo 7 returned to Earth, Soyuz-3 soared into orbit with cosmonaut Georgi Beregovoi on board. He met the unmanned Soyuz-2 which had been launched a day earlier. During the next four days, Beregovoi put on his own folksy television broadcasts for the viewers back home. The Moon Race was rapidly approaching the finish line.

14 Apollo 8 Orbits the Moon

With the success of Apollo 7 and the awesome flights of the Saturn V Moon Rocket in November 1967 and April 1968, Wernher von Braun's engineers declared the Saturn V ready to carry men on its next flight. "There is even a remote possibility," von Braun hinted, "of a spectacular swing around the Moon."

In truth, that possibility was not so remote. The flight of the unmanned Soviet Zond-5 spacecraft around the Moon and back in September 1968, followed by the success of Soyuz-3 in October, led senior NASA officials to conclude that the Russians

On December 21, 1968, a giant Saturn V Moon Rocket launched the Apollo 8 astronauts on mankind's first journey beyond the bonds of Earth's gravity. (NASA)

The Apollo 8 crew extensively photographed the lunar surface, including potential landing sites for future Apollo missions, during their 20 hour, 10 orbit mission around the Moon. This view looks northwest into the Sea of Tranquillity. The lower of the two linear features is the Cauchy Scarp; the upper is the Cauchy Rille. The prominent crater, Cauchy, lies between them. (NASA)

were planning a manned circumlunar mission. While they did not appear ready to actually land on the Moon, a lunar fly-by would still score a major propaganda coup. In November, Zond-6 made another round-trip to the Moon, and the Soviets revealed that Zond was actually "an automatic version of a manned lunar spaceship." American intelligence experts had reason to believe that an attempt to send two cosmonauts to the Moon's vicinity, using a powerful Proton rocket, might be made as early as December.

If the Soviets were not ready to actually land a man on the Moon, neither were the Americans. Lunar Module (LM) deliveries were months behind schedule. But the Apollo 8 crew, which had been training for the first manned test of the LM in Earth orbit, was given an exciting new mission—the Moon!

Faced with the prospect of the Soviets reaching the Moon on a manned mission in the very near future, the bold decision to send Apollo 8 into lunar orbit on Christmas Eve was announced on November 12, 1968. This plan made sense, as Apollo 8's LM would not be ready for launch until March. By sending Apollo 8 around the Moon, NASA would gain valuable time and experience in lunar flight operations before giving

Among Apollo 8's most memorable images were those of Earth, and none are more striking than this view of the "Home Planet" rising above the lunar horizon. The U.S. Postal Service used this photograph on a stamp issued to commemorate the mission. (NASA)

the first LM mission to Apollo 9. While blazing the trail for a landing mission, Apollo 8 would:

- Test the deep space capability of the Manned Space Flight Network for communications and tracking.
- Check Apollo's navigation systems.
- Provide data on the gravitational anomalies discovered by Lunar Orbiter.
- Photograph potential landing sites.

Commander Frank Borman, Command Module Pilot James Lovell, and (LM–less) Lunar Module Pilot William Anders took off from Pad A at Launch Complex 39 on December 21, 1968. Their Saturn V thundered into the sky at 7:51 A.M. EST, arriving in Earth orbit 11 minutes later. The Moonship's systems functioned perfectly, and some three hours into the mission, the astronauts heard good news from Mission Control in Houston: "Apollo 8. You are go for TLI." That was the clearance they needed to re-ignite their S-IVB third stage for translunar injection — the burn that would cut their bonds with Earth, sending them on their way to the Moon.

On December 24, Borman, Lovell, and Anders approached the Moon. As they swung behind the far side, which took them out of contact with Earth, they fired their Service Propulsion System for Lunar Orbit Insertion. The braking action of the rocket engine made them captives of the Moon's gravity. Another Service Propulsion System burn adjusted their orbit to a nearly perfect circle, some 70 miles (114 km) above the lunar surface.

Once again in communication with Earth, Jim Lovell commented on the view below, "The Moon is essentially gray, no color…looks like plaster of Paris or some sort of grayish beach sand."

Bill Anders added, "All you really need is black and white [film]." In a memorable Christmas Eve broadcast, hundreds of millions of viewers saw black and white television images of the lunar surface. "For all the people back on Earth," Anders said near the end of the broadcast, "the crew of Apollo 8 has a message that we would like to send to you."

The astronauts then took turns reading the opening verses from the Book of Genesis: "In the beginning, God created the heaven and the earth…"

Finally, Frank Borman signed off saying, "Goodnight, good luck, a Merry Christmas, and God bless all of you — all of you on the good Earth."

To get back to the "good Earth" — which Lovell described as "a grand oasis in the big vastness of space" — the Service Propulsion System had to fire once again. If it failed, there was no back-up, and the astronauts would be marooned in lunar orbit. The 203 second burn was scheduled to take place early Christmas morning on the far side of the Moon. As the spacecraft emerged from behind the lunar disk, Lovell radioed a welcome report, "Please be informed there is a Santa Claus!" After 20 hours and 10 orbits around the Moon, Apollo 8 was on its way home. Two and a half days later, the astronauts splashed down in the Pacific where they were recovered by the USS Yorktown. Apollo 8 was an unqualified success, and Borman, Lovell, and Anders were hailed as heroes. Addressing a Joint Meeting of Congress on January 9, 1969, Borman said, "Exploration really is the essence of the human spirit. And to pause, to falter, to turn our back on the quest for knowledge, is to perish."

The Apollo 8 mission emphasized engineering more than science, although the astronauts did take some valuable photographs of the Moon, especially the far side. Perhaps some of the greatest returns from Apollo 8 were the images of Earth captured on film and television. During their broadcasts on the way to and from lunar orbit, the astronauts gave us our first clear views of Earth as a planet, suspended in the vastness of space. The poet Archibald MacLeish put those images into words:

> *To see the Earth as it truly is, small and blue and beautiful in that eternal silence where it floats, is to see ourselves as riders on the Earth together, brothers on that bright loveliness in the eternal cold — brothers who know now they are truly brothers.*

Apollo 8 brought us closer to the first lunar landing. But at a time when American society was torn apart by an unpopular war, racial tension, and a generation gap, it also served to bring Americans closer to one another.

15 Reconnaissance Complete!

Apollo 9, which for the first time would test the hardware needed to send two astronauts to the lunar surface, was NASA's most challenging mission to date. Yet, as far as the general public was concerned, it went almost unnoticed, coming as it did after Apollo 8's triumphant Christmas voyage around the Moon and because the spacecraft would remain in Earth orbit. But despite their proximity to Earth, the Apollo 9 crew faced new hazards, one of which was flying in a vehicle that could not re-enter Earth's atmosphere in case of trouble.

Commander James McDivitt was joined by Command Module Pilot David Scott, and Lunar Module Pilot Russell "Rusty" Schweickart. Lifting off Pad 39A on March 3, 1969, the astronauts rode a Saturn V into orbit for a 10-day mission that would give program managers even more confidence in Apollo. Three hours after launch, Scott pulled the Command/Service Module (CSM) away from the third stage, turned around, and docked with the Lunar Module (LM), which was still attached to the S-IVB. Explosive bolts and compressed springs gently pushed the combined vehicles away from the rocket. Because the CSM and LM perform maneuvers separately from one another, call signs would be required throughout the remainder of the Apollo Program to denote the CSM and the LM. In the case of Apollo 9, the astronauts chose *Gumdrop* for the CM and *Spider* for the LM. The names seemed to fit, given the shapes of the two ships.

During the first three days of the mission, the crew kept busy checking equipment, firing their big Service Propulsion System engine, and preparing for the LM's test program. On Day Two, McDivitt and Schweickart floated into *Spider* and powered up its systems. During that procedure, Schweickart became sick — vomiting on two occasions. This was not unusual, as many astronauts before him had experienced a similar problem during their first day or two in weightlessness. However, it did cause some concern with

Three hours after the launch of Apollo 9, the Command/Service Module Gumdrop *pulled away from the third stage of the SIV-B, turned around, and docked with the Lunar Module* Spider. *Here, in a photograph taken from* Gumdrop, *the Lunar Module is seen still attached to the S-IVB as the Command/Service Module approaches* Spider's *circular docking port. On trips to the Moon, this "transposition" maneuver would be completed after the translunar injection burn sent the astronauts Moonward.* (NASA)

regard to Schweickart's planned spacewalk on the fourth day of the mission. This was a very important spacewalk, as it was designed to test the new Apollo Extravehicular Mobility Unit (EMU)—a Moonsuit which in many respects could be considered a third spacecraft. The EMU was equipped with a Portable Life Support System (PLSS, or backpack) which provided the oxygen, thermal control, and communications equipment required for extended excursions on the lunar surface.

Because of his illness, Schweickart's spacewalk was somewhat abbreviated. But he crawled through *Spider's* forward hatch onto the front porch where he enjoyed a spectacular view of Earth, as well as of Dave Scott, who was standing in *Gumdrop's* open hatch. The EMU worked perfectly.

Apollo 9's fifth day marked the main event. McDivitt and Schweickart entered *Spider,* powered up its systems once again, and separated from Dave Scott in *Gumdrop.* During the next six hours, McDivitt and Schweickart flew the LM through a series of complex maneuvers, designed to test *Spider's* rendezvous radar and computer, as well as its descent and ascent engines.

Apollo 10's Command/Service Module Charlie Brown *is seen in this head-on view taken from the Lunar Module* Snoopy. *The two spacecraft are circling 70 miles (114 km) above the lunar surface.* (NASA)

The LM passed its first manned test with flying colors, and the astronauts were safely reunited. After spending five more days in orbit performing a variety of experiments, the Apollo 9 astronauts splashed down in the Atlantic on March 13, 1969. The Earth orbital tests had reached a successful conclusion, but NASA decided to make one more test of the entire system before attempting to land the first men on the Moon.

On May 18, 1969, Apollo 10 rumbled away from Pad 39B. It would come within 50,000 feet (15,250 m) of the lunar surface in a dress rehearsal for Apollo 11. Space veterans Commander Thomas Stafford, Command Module Pilot John Young, and Lunar Module Pilot Eugene Cernan made up Apollo 10's crew. They carried a color television camera, providing spectacular views of Earth on their way to the Moon. Three days and a quarter of a million miles after lift-off, Apollo 10 went into lunar orbit.

Astrogeologists were intrigued by Stafford's report of seeing a "couple of good volcanoes" which Young further described as "all white on the outside, but definitely black inside." These remarks fueled the ongoing controversy between those who believed volcanoes had formed the Moon's craters and those who advocated the impact theory of formation, even though Surveyor and Lunar Orbiter had already essentially proven that most lunar craters had been blasted out by incoming space debris such as meteorites, comets, and asteroids.

Apollo 10's mission was not to settle scientific arguments, but rather to test the Apollo Earth-Moon transportation system and to practice Lunar Orbit Rendezvous. With those goals in mind, Stafford and Cernan entered the LM *Snoopy* on May 22, and powered up its systems while Young kept watch in *Charlie Brown,* the call signs being taken from the popular *Peanuts* comic strip by Charles Schulz. After verifying that the two spacecraft were functioning properly, *Snoopy* undocked from *Charlie Brown,* and Stafford and Cernan started their descent to within 9½ miles (15.3 km) of the lunar surface. The descent engine worked perfectly, and the astronauts reported that the primary landing site, dubbed Apollo Landing Site 2 (ALS 2), and one of five sites deemed suitable for an early mission, appeared "pretty smooth, like wet clay, like a dry river bed in New Mexico or Arizona."

Just before they jettisoned the Descent Stage and fired their ascent engine to climb back to the CSM, Stafford and Cernan experienced a wild, eight-second gyration caused by a malfunction in the backup guidance system. The LM's erratic behavior created a brief scare, but Stafford quickly regained control, and *Snoopy* was reunited with *Charlie Brown* after an eight hour solo flight that skimmed the Moon's oceans and mountain ranges.

The trip home was uneventful, and Apollo 10 splashed down in the Pacific on May 26, 1969. The reconnaissance of the Moon was complete.

The stage was now set for Apollo 11 to fulfill a national goal. And a centuries old dream.

16 *"The Eagle has landed."*

As dawn broke over the Florida coast on July 16, 1969, an estimated one million people were gathered at and around Cape Canaveral to witness the historic launch of Apollo 11, the mission that would carry the first humans to the surface of another world. Among them were Wernher von Braun and his 75-year old mentor, Hermann Oberth.

The first moments in the flight of Apollo 11 were captured by a wide-angle camera mounted atop the umbilical tower at Launch Pad 39A. Clearly seen atop the rocket are the Launch Escape System and the Boost Protective Cover which shielded the Command Module. (NASA)

The crew for the historic Apollo 11 mission included (left to right): Commander Neil A. Armstrong, Command Module Pilot Michael Collins, and Lunar Module Pilot Edwin E. ("Buzz") Aldrin, Jr. (NASA)

Like the spectators, the astronauts had risen early for this epoch-making event. Commander Neil A. Armstrong, Command Module Pilot Michael Collins, and Lunar Module Pilot Edwin E. "Buzz" Aldrin, were awakened at 4:15 A.M. EDT. After a hearty breakfast, they suited up and headed out to Pad 39A. Armstrong was the first to board the Apollo 11 Moonship, taking his place in the left-hand seat. Collins took his place on the right, with Aldrin in the middle (Lunar Module pilots usually took the right-hand seat). At 9:32 A.M. the big Saturn V roared to life, hurling Apollo 11 into space.

Early that afternoon, after leaving Earth orbit, Collins took control of the Command/Service Module *Columbia,* and pulled ahead of the S-IVB. He turned around, docked with the Lunar Module (LM) *Eagle,* and extracted the lander from the third stage to which it had been attached.

Two days later, while 48,000 miles (77,000 km) from the Moon, the astronauts transmitted a televised tour of the LM, giving viewers on Earth a look at the interior of the ship that would carry Armstrong and Aldrin to the lunar surface. The following afternoon, July 19, the astronauts fired *Columbia's* Service Propulsion System engine, braking them into lunar orbit. Describing his first view of the landing site, Armstrong said, "It looks very much like the pictures, but like the difference between watching a real football game and watching it on TV—no substitute for actually being here."

Apollo Landing Site 2, one of three sites considered for the first landing, was a relatively smooth, mare surface, chosen for safety and operational considerations. It was located near the southwestern edge of the Sea of Tranquillity, close to the lunar equator and just to the right of the center of the Moon (as viewed from Earth). The nearest highlands were 25 miles (41 km) to the south, and the Surveyor 5 landing site was just 15 miles (25 km) to the northwest.

One factor affecting the timing of the landing was the position of the Sun. Arriving near the Moon's terminator shortly after local sunrise meant that long shadows would give the astronauts some measure of topographic relief, making it easier to see and avoid potential hazards. The approach was to be from the east (all Apollo spacecraft orbited the Moon from east to west). If communications with Earth were to be maintained during the critical, final descent phase, lunar touchdown sites could not be too close to the Moon's eastern limb (the right edge as seen from Earth) because the landing would be initiated while still on the far side.

On the morning of July 20, 1969, Armstrong and Aldrin entered the LM once again and powered up its various systems. A few hours later, *Eagle* separated from *Columbia*. "The Eagle has wings!" Aldrin exclaimed.

At 3:08 P.M. EST, Armstrong fired the LM's descent engine, lowering *Eagle's* orbit to 50,000 feet (15,250 m). At that point, with the crew flying feet-first and face-down relative to their flight path, the powered descent initiation maneuver began, braking the LM out of lunar orbit and putting it on course toward a touchdown, 300 miles (480 km) downrange. As *Eagle* dropped lower and lower, it assumed a more vertical attitude. The spacecraft rolled around to give its crew a better view of the approaching landing site and Armstrong saw that the LM was going to overshoot its target. More importantly, he saw that *Eagle's* autopilot was taking him into a boulder-strewn, football field–sized crater. Assuming manual control, Armstrong used up precious fuel as he steered the LM to a smoother site.

Aldrin called out the craft's altitude. "Four hundred feet, things looking good… Lights on… Picking up some dust… 30 feet, two and a half down… Faint shadow… Four forward. Four forward, drifting to the right a little…"

At this point, one of the 68 inch (1.7 m) probes that was attached to three of *Eagle's* four footpads, made contact with the surface, signaling the crew to shut down the engine. "Contact light," Aldrin said, calmly, almost routinely. "Okay, engine stop."

At 4:18 P.M. EDT, Armstrong radioed Mission Control, "Houston… Tranquillity Base here. The Eagle has landed."

A little over six and a half hours later, the two men were ready to explore the surface. They depressurized the cabin and opened the front hatch. With an 84 pound (38 kg) Portable Life Support System strapped to his back—it only weighed 14 pounds (6.4 kg) on the Moon—Armstrong slowly made his way through the door and onto the LM's

Buzz Aldrin clambered down Eagle's *ladder 15 minutes after Neil Armstrong had stepped onto the lunar surface. Here he backs off the "porch" and onto the first rung of the ladder, having closed the Lunar Module's forward hatch behind him.* (NASA)

porch. He pulled a D-ring which deployed a television camera to record his first step on the Moon.

"I'm at the foot of the ladder..." he announced. "The surface appears to be very, very fine-grained as you get close to it. It's almost like a powder." At 10:56 P.M., Armstrong stepped off the LM footpad, sticking his left boot into the powdery lunar soil and said, *"That's one small step for man, one giant leap for mankind."*

Millions of viewers on Earth shared that unforgettable moment with Armstrong via ghostly black and white television images. Describing the lunar landscape at Tranquillity Base, Armstrong said, "It has a stark beauty all its own." A few minutes later, Aldrin joined Armstrong on the surface. Aldrin remarked on the Moon's "magnificent desolation."

With less than 2½ hours allocated for their Moonwalk, the astronauts had little time to play the role of tourists. They did however, work a few minutes into their schedule to raise the Stars and Stripes, unveil a commemorative plaque, and accept a congratulatory telephone call from the President.

Most of the astronauts' time was dedicated to collecting rock and soil samples, taking photographs, and setting up three experiments: a Passive Seismic Experiments Package (PSEP) to record and measure Moonquakes and meteorite impacts; a Laser Ranging Retro-Reflector (LRRR or LR^3) to precisely measure Earth-Moon distances; and a Solar Wind Experiment, which was in essence a sheet of aluminum foil designed to collect particles of solar wind which blow continuously outward through space from the Sun. These particles are deflected past our planet by Earth's strong magnetic field, but the Moon's weak magnetic field allows them to hit the lunar surface. Bits of the Sun would therefore be retrieved on the Moon!

The PSEP was a lunar seismometer which functioned for only a few weeks, long enough to suggest that the Moon was essentially a dead world with only an occasional small meteorite or minor Moonquake to shake its eternal stillness.

The LR^3 was comprised of 100 optical reflectors. It was used to reflect a laser beam sent from Earth back to Earth. By measuring the time for the beam's round trip, Earth-Moon distances could be determined to an accuracy within inches or centimeters, enabling scientists to precisely study the motion of the Moon and to measure variations in Earth's rotation.

The Solar Wind Experiment was referred to as the Swiss Flag because it was sponsored by the Swiss government and it looked like a flag. Mounted on a staff, it rolled up and down like a window shade. After the Swiss Flag was exposed to the solar wind during their Moonwalk, the astronauts rolled it back up. Scientists on Earth later identified the particles ejected by the Sun as mostly atoms of hydrogen and helium, the Sun's main constituents.

With the experiments deployed, Armstrong and Aldrin returned to the LM with 48.5 pounds (22 kg) of carefully selected Moon rocks and soil samples, and hundreds of photographs that documented their experiences on the Moon.

Armstrong and Aldrin rested in the LM that night. The following afternoon, they fired *Eagle's* ascent engine and redocked with *Columbia* for the voyage home. Apollo 11 splashed down in the Pacific on July 24, and the astronauts were welcomed aboard the *USS Hornet* by President Nixon. The crew was quarantined for the next three weeks against the unlikely possibility of their contaminating Earth with Moon germs (there were none). Quarantined with them at the Lunar Receiving Laboratory in Houston, were the first rocks ever retrieved from another planet—the most valuable rocks on Earth!

On September 15, 1969, NASA held a press conference in Washington to present a preliminary analysis of the lunar rocks. Most of the Apollo 11 samples were basalts formed by lava at or near the surface. They contained some of the same minerals found on Earth, though in somewhat different proportions. They contained absolutely no water. The rocks confirmed the findings of Surveyor 5, and confirmed that impacts had played a very important role in the evolution of the Moon. Still, in order to gain a clearer understanding of the Moon's history—as well as that of Earth—more samples would be needed from more diverse regions of the Moon.

17 Lunar Wardrobe

Watching Neil Armstrong and Buzz Aldrin hop, bounce, and glide across the lunar surface, television viewers on Earth could see the effects of the Moon's reduced gravity. It looked like fun but less obvious were the many extreme hazards and difficulties associated with Moonwalking.

Whether on the Moon or in orbit around Earth, Extravehicular Activity (EVA) exposes an astronaut to the near vacuum of space. Without protection, the gases in the body rapidly expand, the blood appears to boil, and a person dies within seconds. But the astronauts were quite safe inside their multi-layered, airtight, pressure suits, though the stiff, bulky garments did greatly reduce their mobility. The closest most of us come to experiencing this is the awkwardness of wearing several layers of clothing on a cold winter day.

The astronauts faced other dangers on the Moon. Here on Earth, where thick layers of atmosphere block most of the Sun's harmful rays, we can still feel the strong effects of solar radiation, especially at the beach on a hot summer day. On the Moon, daytime temperatures climb to well above the boiling point of water, and the lunar soil is hotter than the hottest beach sand. Even wearing lunar sandals, an astronaut would burn his feet. Likewise, infrared and ultraviolet radiation from the Sun would permanently damage his eyes. What a sunburn!

The Moonwalkers also faced the remote possibility of a meteoroid penetrating their suits, either killing them as a bullet would, or making a hole, through which their oxygen would escape. There was also the threat of a tear, caused by a fall on a jagged rock or sharp piece of equipment.

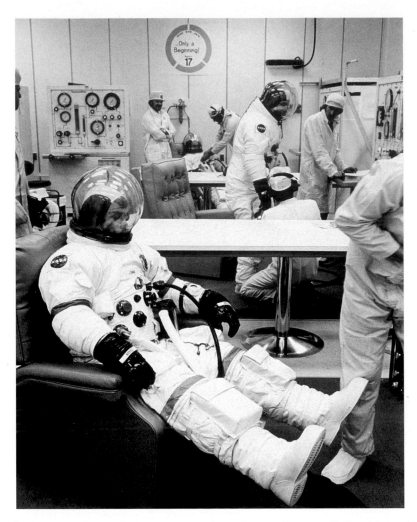

Suiting up for the Moon, the Apollo 17 astronauts, Gene Cernan (foreground), Ron Evans (middle) and Jack Schmitt are helped into their bulky pressure suits by NASA technicians. The task was easier in space where the men helped each other don and remove their individually tailored suits. (NASA)

The Moon presented a tough challenge for spacesuit designers, but the challenge was well-met with an ensemble called the Extravehicular Mobility Unit (EMU). The Apollo EMU incorporated five major components:

- **A liquid-cooled undergarment** replaced the one-piece undergarment (similar to longjohns) and coveralls that the astronauts wore inside the spacecraft. This liquid-cooled undergarment was made of knitted, nylon-spandex and had a 265 foot (81 m) network of plastic tubing, through which cooling water from the Portable Life Support System circulated. The water carried away the astronauts' metabolic (body) heat. The system could be regulated to allow for differing levels of activity.

A spacesuit, the self-contained personal spacecraft called an "Extravehicular Mobility Unit" by NASA, is seen in this head-on view of Buzz Aldrin taken by Neil Armstrong during the Apollo 11 mission. The box-like device on Aldrin's chest contained controls for the Portable Life Support System strapped to his back. The coated visor afforded protection from the Sun. (NASA)

- **An outer pressure garment** (spacesuit) included a helmet and lunar gloves. The pressure garment worn over the liquid-cooled undergarment provided thermal and meteoroid protection. The rubber-coated, inner layer had specially designed shoulder, elbow, wrist, waist, knee, and ankle joints that gave the astronauts limited ability to bend and reach while working on the Moon. The suit included 18 layers of material, with the outer shell being an abrasion-resistant layer of white, Teflon fabric. A fishbowl helmet fit over the head and offered freedom of movement, but the astronauts still couldn't rub their noses! Caps that carried microphones and earphones were worn under the helmets, and provision was made on later flights for a quart bag of drinking water to be attached inside the helmet. EVA gloves gave

extra protection for handling extremely hot or cold objects. The fingertips were
made of silicon rubber to provide more sensitivity. Special lunar boots were worn
to protect the soles of the suit from heat and rocks.

- **An EVA visor assembly,** consisting of a polycarbonate shell and two visors with
  thermal control and special coatings, fit over the helmet to provide impact, mi-
  crometeoroid, thermal, and ultraviolet and infrared ray protection to the
  Moonwalker. On later missions, a sunshade was added to the outer portion of the
  EVA visor assembly.

- **A Portable Life Support System (PLSS)** was an amazing backpack that
  contained oxygen, cooling water, and communications equipment. In addition, it
  absorbed carbon dioxide and removed excess humidity from the astronaut's air
  supply. The PLSS was equipped with tanks, pumps, and batteries, plus a control box
  worn on the chest.

- **An Oxygen Purge System (OPS)** was mounted atop the PLSS. The OPS
  provided a 30 to 75-minute emergency air supply.

The EMUs on the last three Apollo missions enabled an astronaut to remain outside
the Lunar Module for up to seven hours at a time and could be recharged for three
separate EVAs.

A Moon walker in an EMU was, in effect, the pilot of a self-contained spacecraft!

18 Science on the Moon

The first Moon landing was a historic achievement, as well as a symbol of American
technological supremacy. Although it returned the first rocks from another world,
Apollo 11 was more an engineering demonstration than a scientific expedition. The
mission proved that the basic spacecraft design was sound and that man could live and
work on the lunar surface.

Apollo 12 would further test the elaborate system for manned lunar landings, but it
would also devote more time and attention to science. Launched into a threatening sky
on November 14, 1969, the mission was nearly aborted shortly after lift-off when a
lightning bolt temporarily knocked out the spacecraft's electrical system. Commander
Charles "Pete" Conrad, Command Module Pilot Richard Gordon, and Lunar Module
Pilot Alan Bean suddenly saw warning lights illuminating their instrument panel. Flight
controllers quickly assessed the situation. A decision was made to continue the ascent,
and as the crew reset circuit breakers and switches, the system came back on line. Addi-

tional checks were made while in Earth orbit, and everything functioned normally. Apollo 12 was given a go for translunar injection.

The all-Navy crew had named their Command/Service Module *Yankee Clipper* and their Lunar Module (LM) *Intrepid* after famous sailing ships. *Intrepid* was targeted for Apollo Landing Site 7, which was 1,300 miles (2,090 km) west of Tranquillity Base in the Moon's Ocean of Storms. A ray of ejecta material, blasted from Copernicus, located 250 miles (400 km) to the north, had crossed the mare plain at Landing Site 7. Conrad and Bean might have an opportunity to bring back pieces from that giant crater.

One of the most dangerous tasks assigned to the Apollo 12 astronauts was to extract a radioactive fuel rod from a graphite cask mounted near the rear of the Lunar Module descent stage. Here Alan Bean cautiously pulls the rod from its cask, using a special tool. He then inserted it into the SNAP-27 nuclear generator seen to the right of his knees. This device powered the Apollo Lunar Surface Experiments Package (ALSEP) left behind on the lunar surface. (NASA)

If their landing was close enough to the target, they would also have a chance to bring back parts from Surveyor 3, which had touched down at the site some 31 months earlier. Scientists wanted to study the effects of long-term exposure to the lunar environment on assorted spacecraft materials by retrieving Surveyor 3's television camera and other bits.

One of two unusual mounds observed by Pete Conrad and Alan Bean during their first Moonwalk. These volcano-shaped features were probably just piles of coarse material thrown out of a crater by a meteorite impact. (NASA)

Whereas the Early Apollo Scientific Experiments Package (EASEP) left behind on the Moon by Apollo 11 had a seismometer powered by solar panels, Apollo 12 carried a nuclear-powered Apollo Lunar Surface Experiments Package (ALSEP), which was the first of an eventual network of five long-term, unmanned scientific stations deployed on the Moon. Each ALSEP contained several automatic instruments designed to transmit a continuous flow of readings and data over a period of months or years (see Chapter 19 for a more detailed description of the ALSEP).

Alan Bean holds a Special Environmental Container in which a carefully selected lunar sample is preserved in a high-vacuum environment. It was later opened at the Lunar Receiving Laboratory in Houston. (NASA)

Pete Conrad brought *Intrepid* down near the west rim of a 650 foot (200 m) crater at 1:54 A.M. EST on November 19, 1969. Five hours later, he and Al Bean started the first of two planned lunar excursions. As Conrad set foot on the Moon, he spotted Surveyor 3 sitting inside the crater, just 535 feet (163 m) east of the LM. With the help of NASA computer experts on Earth, Conrad had made a remarkably accurate touchdown. "Old Surveyor," as Conrad called it, had visitors!

Surveyor 3 bounced twice when it landed—leaving multiple footprints in the Moon's soil. Despite 31 months of exposure to the lunar environment, those imprints had not been altered—nor had the small pile of dirt dumped on Surveyor's footpad by its remotely-controlled scoop. (NASA)

Before the astronauts could make their way down the crater wall to inspect the unmanned scout, they had more important work to accomplish: gathering rock samples and setting up the ALSEP station. Television viewers back home had hoped to follow the astronauts' progress, this time in color. Unfortunately, shortly after Bean set foot on the surface, he inadvertently pointed the camera at the Sun, damaging the camera's vidicon tube. The television pictures were lost, but the two astronauts still provided a colorful — sometimes giddy — audio account of their activities.

The first thing they did was set up an umbrella-like S-band antenna to augment their radio signal strength for improved communications with Earth. Next, they planted an American flag. Then they unloaded the ALSEP from a storage compartment in the Descent Stage. One of their most dangerous tasks was to extract a 932 degree Fahrenheit (500 degree Centigrade) plutonium-238 fuel capsule from its protective graphite cask.

This radioactive rod was then inserted into a SNAP-27 nuclear generator which sup-plied electrical power to the ALSEP experiments. That job accomplished, Conrad and Bean chose a site 425 feet (130 m) northwest of the LM to deploy the instruments. By the time they got there, both men were covered with the powdery, black, lunar soil that plagued every Apollo landing team as they explored the Moon's surface.

Apollo 12's ALSEP included: a seismometer, a surface magnetometer, an atmosphere detector, an ionosphere detector, and a solar wind spectrometer. A solar wind composi-tion experiment — another Swiss Flag — was also set up by the astronauts near the LM. While it would be brought back to Earth by Conrad and Bean, the ALSEP instruments would stay behind to send back valuable new data about the lunar environment.

Throughout their first Moonwalk, Conrad and Bean described the surrounding landscape, including two small mounds. These volcano-shaped features were probably piles of coarse material thrown out of a crater. The two astronauts also collected rock samples, carefully noting the location of each and photographing its orientation on the surface. This information would enable geologists to better understand each sample's history, for example, whether it had come from a specific crater. In addition, the astro-nauts hammered several core tubes into the surface of the Moon. These pipe-like, alumi-num tubes were driven nearly three feet (one meter) into the ground to obtain a cross-section of the subsurface, which might reveal layering in the regolith.

After nearly four hours outside the LM, the astronauts dusted each other off as best they could, stowed their rocks and core samples, and climbed back into *Intrepid* for dinner and a night's rest. Hammocks were strung up in the cabin to make sleeping in the confined quarters of the LM a bit more comfortable.

Some 12 hours after their first Moonwalk had ended, Conrad and Bean were ready to begin their second four-hour Moonwalk. This time, the focus was geology, and their pre-planned route took them past a variety of craters. They picked up representative rocks along the way, and scooped up some light gray material, later identified as the ejecta from Copernicus, they had hoped to find. The seismometer at the ALSEP station registered their footsteps, as well as the movement of a rock that Conrad rolled down the side of one crater wall. The astronauts also sampled the soil at a small, relatively fresh rayed crater before making their way to Surveyor Crater. Crossing over its southern edge, they followed a curving path, parallel to the rim, finally reaching Surveyor 3. The men photographed the surrounding area for comparison with the images taken by Sur-veyor to see if any changes had taken place since 1967 (none had). Then they cut off Surveyor 3's camera, a few pieces of tubing and cable, and the scoop. These treasures in hand, they headed back to *Intrepid* and prepared for their departure the next morning.

After docking with *Yankee Clipper*, the LM ascent stage (the mass and velocity of which were known) was sent crashing into the Moon to calibrate the ALSEP seismom-eter. The result was surprising. The Moon rang like a bell for nearly an hour, suggesting that shock waves from the impact bounced through a severely fragmented structure sandwiched between two reflective layers of rock, extending roughly six miles (10 km) beneath the surface.

Apollo 12 splashed down in the Pacific on November 24, 1969. The 75 pounds (34 kg) of samples brought back from the Ocean of Storms were mostly volcanic basalts, like those returned by Apollo 11, but some suggested the Moon had a complex past. While Apollo was resolving some old questions, new ones were being raised.

19 Apollo Lunar Surface Experiments Package (ALSEP)

From the very beginning of the Apollo program, a scientific instrumentation system to study the Moon was included in the Lunar Module (LM) design specifications, although the exact nature of the system was undefined. During the summer of 1965, scientists, representing a number of disciplines, gathered for a conference in Falmouth, Massachusetts to identify the types of experiments that should be conducted on the Moon. A modular system, using a central station for processing and transmitting data, as well as receiving commands from Earth, would allow different instruments to be interchanged for specific missions. Each instrument would be connected with the central station by flat, ribbon-like electrical cables, and the entire system would be powered by a Radioisotope Thermoelectric Generator (RTG), delivering roughly 70 watts. The RTG was designed to operate for at least one year.

Early in 1966, NASA selected the Bendix Corporation to build the central station and to integrate the various instruments in the Apollo Lunar Surface Experiments Package (ALSEP). The SNAP-27 RTG was developed by the Atomic Energy Commission under its SNAP (Systems for Nuclear Auxiliary Power) program. The spontaneous, radioactive decay of the RTG's plutonium-238 fuel, generated heat, and an assembly of 442 lead telluride thermocouples converted the temperature difference between the fuel capsule and the finned outer housing directly into electrical energy. There were no moving parts in the generator.

The central station, instruments, and RTG combination could weigh no more than 150 pounds (68 kg) and take up no more than 12 cubic feet (0.34 cu m) of space. The ALSEP was housed in a rear compartment of the LM descent stage to give the astronauts easy access to it, after landing. Mounted on the LM was a separate graphite cask for the plutonium-238 fuel capsule. It was designed to withstand Earth re-entry and impact in case of an aborted mission (as eventually occurred during the flight of Apollo 13). The entire ALSEP package fit neatly into a pallet, the two halves of which could be easily carried by an astronaut using a barbell-like handle.

Although the ALSEP package was small, its instruments were extremely sensitive and promised to provide a great deal of information about the lunar environment. Seven experiments were selected for inclusion in the ALSEP:

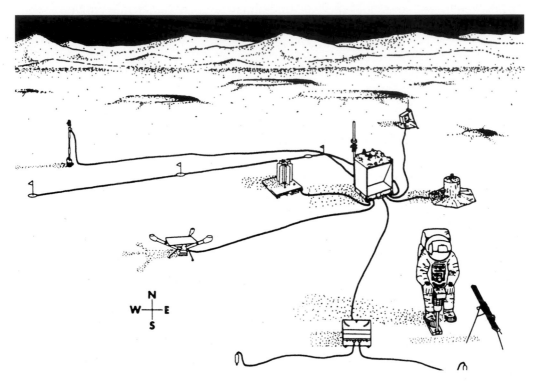

A typical Apollo Lunar Surface Experiments Package (ALSEP) layout is depicted in this diagram of the Apollo 16 station. The experiments were connected by cables to a central station (the box-shaped structure above and to the left of the astronaut who is using a lunar surface drill to obtain a core sample) which received commands from Earth and processed and transmitted data back to Earth. Note the Radioisotopic Thermoelectric Generator (RTG) to the left of the central station. (NASA)

1. The **Passive Lunar Seismic Experiment**, a three-axis seismometer, was designed to record meteorite impacts and to measure Moonquakes in the Moon's interior.

2. The **Active Lunar Seismic Experiment** was designed to study the physical properties of the lunar regolith to a depth of roughly 1,000 feet (300 m) by using three geophones and a thumping device (carried by an astronaut) to measure the propagation of sound waves through the soil. A small mortar, firing grenades, would also be employed after the astronauts had left the Moon. On Earth, this technique is used by geologists to find new oil and gas fields, but on the Moon it was looking for a layer of solid rock—if one existed—near the lunar surface.

3. The **Lunar Tri-Axis Magnetometer** was designed to measure the weak lunar magnetic field to help determine if the Moon has, or once had, a molten core like Earth.

4. The **Medium Energy Solar Wind Experiment** was designed to measure the velocity and direction of protons, electrons, and alpha particles radiated by the Sun, and their interaction with the lunar surface.

5. The **Low Energy Solar Wind Experiment** was designed to study solar wind particles.

6. The **Suprathermal Ion Detector** was designed to measure the Moon's tenuous atmosphere.

7. The **Lunar Heat Flow Measurements** was designed to sink probes into the regolith to measure the outflow of heat from the Moon's interior to its surface, and to provide information on the distribution of the Moon's radioactive elements as well as its thermal history.

Several additional experiments were eventually included as ALSEP instruments, such as the **Lunar Surface Gravimeter** which searched for gravitational waves in space, and the **Lunar Ejecta and Meteorite Experiment** to measure the speed, direction of travel, and mass of micrometeorites hitting the surface of the Moon, as well as similar data relating to any lunar particles that were ejected from the Moon by large meteorite

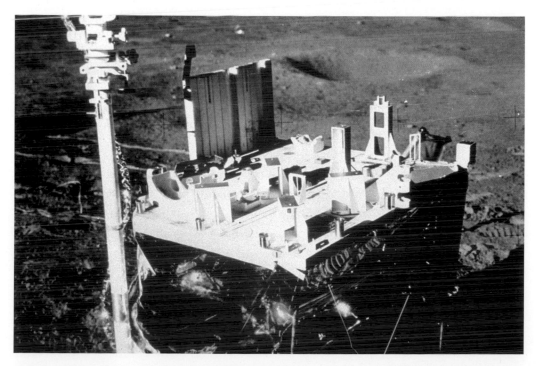

A close-up of the Apollo 14 ALSEP (Apollo Lunar Surface Experiments Package) central station and the base of its helical S-band antenna. The central station and Radioisotopic Thermoelectric Generator (RTG) were common to all five ALSEP stations, but some of the experiments connected to them were different. (NASA)

The Lunar Tri-Axis Magnetometer can be seen in the foreground of this view showing the Apollo 12 ALSEP. The astronaut in the background is standing by the central station with its pole-like antenna pointing toward Earth. Additional experiments ring the rim of a shallow crater. (NASA)

impacts. These experiments allowed scientists to study the Moon long after the astronauts had returned to Earth.

Had all of the originally scheduled Moon flights taken place, and had Apollo 13 been successful, nine ALSEP stations would have sent data from a variety of sites widely scattered across the lunar surface. However, in January 1970, NASA announced the cancellation of Apollo 20 in order to provide funding and hardware for the Skylab program, an Earth-orbiting space station to be launched by the last of 15 Saturn V rockets produced for Apollo. Nine months later, two more Apollo lunar missions were canceled. The American public had lost interest in the Moon, and the near disaster experienced by Apollo 13 in April of 1970 had eroded political support for space exploration. Consequently, only five ALSEP stations were deployed on the lunar surface. These stations transmitted useful data long after their planned operational lifetimes. Since each ALSEP could send some nine million instrument readings a day, the quintet returned a truly staggering amount of information.

Then, on September 30, 1977, another cost-cutting measure was placed into effect. Although the ALSEP stations were still sending signals to Earth, NASA stopped receiving them, shutting down the first network of scientific instruments established on another world, in order to save $200,000 a year. As a result, many questions which could have been answered about the Moon remain unanswered. But before they were abandoned, the ALSEP experiments provided scientists with a bonanza of lunar data. The experiments had also proven the value of sending humans to the Moon: without the astronauts, it is doubtful that the ALSEP stations could have been deployed.

20 From the Moon, Knowledge

The deployment of the Apollo Lunar Surface Experiments Package (ALSEP) instruments by the Apollo 12 astronauts opened a new era in lunar science. The rocks and soil samples (referred to as "fines" by geologists) shed new light on an old Moon. As many scientists had theorized, the lunar maria were indeed formed by basaltic lava flows similar to those which are still helping shape Hawaii today, although the molten lava on the Moon was probably more fluid. The lava cooled, was slowly broken up into rocks, and pulverized into soil by meteorite bombardment, creating a fragmented, but firm regolith about 3 to 50 feet (1 to 15 m) thick.

The Apollo 11 rocks were 3.7 billion years old. Those picked up by Conrad and Bean on Apollo 12 were 3.2 billion years old, the lava flows at the Apollo 12 landing site having poured across the Ocean of Storms some 500 million years after those in the Sea of Tranquillity. There were also differences in the chemical composition of the rocks from the two missions. The Apollo 12 samples, for example, contained considerably larger concentrations of elements such as potassium, thorium, and uranium.

Clearly the Moon was not a homogeneous body, which is why Apollo 13 was targeted away from the maria and toward a hilly area called the Fra Mauro Formation

The courageous crew of the ill-fated Apollo 13 (left to right): Lunar Module Pilot Fred W. Haise, Jr.; Command Module Pilot John L. Swigert, Jr.; and Commander James A. Lovell, Jr. (NASA)

The Apollo 13 Command/Service Module is seen here in the Manned Spacecraft Operations Building at the Kennedy Space Center prior to mating with its Saturn V launch vehicle in the Vehicle Assembly Building. (NASA)

(named for a 15th Century Venetian geographer). The landing site was 112 miles (180 km) east of where Apollo 12 had touched down. Although it was adjacent to the crater Fra Mauro, the landing area was actually representative of a much larger portion of the Moon. The reason for this is that the Fra Mauro Formation was thought to be ejecta from the huge impact that blasted out the Imbrium Basin (which, when filled by lava, became Mare Imbrium). The impact melted the Moon's crust and splashed out huge amounts of molten material radially from its center, some of it falling at Fra Mauro. Forming hummocky terrain higher than the surrounding plains, the molten material was never covered by subsequent lava flows. Therefore, geologists believed that samples from Fra Mauro could be used to date the Imbrium event which, in turn, would help them determine the relative ages of other regions on the Moon.

Given the importance of the Fra Mauro landing site to lunar science, the Apollo 13 crew included a motto on their mission patch: *Ex Luna, Scientia* (*From the Moon, Knowl-*

Just hours before splashdown, the Apollo 13 astronauts got their first look at the damaged spaceship when they shed the Service Module in preparation for re-entry. "There's one whole side of that spacecraft missing!" Lovell reported. Even through the Sun's glare, a gaping hole can be seen where panel 4 had previously been. (NASA)

edge). Commander James Lovell and Lunar Module Pilot Fred Haise diligently prepared for their geological traverse across Fra Mauro. Command Module Pilot Thomas "Ken" Mattingly studied lunar geography to help him photograph selected regions of the Moon's surface from lunar orbit.

A few days before launch, Mattingly was exposed to rubella—German measles—and had to be replaced by his back-up, John "Jack" Swigert. Lift-off came at 2:13 P.M. EST on April 11, 1970. After the Command Module (CM) *Odyssey* extracted the Lunar Module (LM) *Aquarius,* from its adapter, the S-IVB was sent on a collision course toward the Moon. The effects of its powerful impact was recorded by the Apollo 12 seismometer.

For 2½ days, the flight proceeded normally. But, as Jim Lovell's 1994 account, *Lost Moon* so dramatically related, an explosion in a liquid oxygen tank crippled the Service Module (SM), canceled the Moon landing, and forced the three astronauts to use the

LM as a lifeboat for their perilous 3½-day journey back to Earth.

Few who lived through the Apollo era will ever forgot Jack Swigert's first report of trouble on the evening of April 13: "Hey, we've got a problem here." Mission Control then asked him to repeat that message, and Jim Lovell responded, "Houston, we've had a problem." The astronauts had heard a loud bang and felt a vibration rock their ship. Outside their windows, they could see a gas rapidly venting into space from their SM. Inside the CM, warning lights were flashing on the instrument panel as the spacecraft's gauges recorded the sudden loss of two out of three critical fuel cells, *Odyssey's* primary source of water and electricity.

An explosion of unknown origin had knocked out the Command/Service Module's electrical power and oxygen supply. With just 15 minutes of life left in *Odyssey,* the astronauts shut down the CM's remaining systems (saving them for Earth re-entry) and transferred to *Aquarius.* The three men would now have to spend the next 90 hours in a cramped spacecraft designed to support two men for 45 hours. And the LM's comparatively small descent engine, rather than the SM engine, would have to be fired (using an unproven technique) to put Lovell, Haise, and Swigert on a return course to Earth.

During their trip, the astronauts were forced to conserve the LM's limited battery power by turning off all but its most essential systems. Without heat, the temperature in the cabin plummeted to just above the freezing point. The men shivered in the cold. Moreover, they had to conserve their water supply, drinking less than a cup a day. As a result, they became dehydrated and lost their appetites.

The astronauts' most life-threatening problem was the air they were breathing. *Aquarius* did not have enough lithium hydroxide canisters — which removed carbon dioxide from the spacecraft — to make it home. There were extra canisters back in *Odyssey,* but the CM's square units did not fit the round holes in the LM's environmental system. As the carbon dioxide rose to dangerous levels, NASA engineers back on Earth worked feverishly to design an improvised device that would enable the astronauts to adapt the CM's canisters to the LM's system. Using a spacesuit hose, cardboard maps, plastic bags, and tape — materials that could be found on board their spacecraft — Lovell and Swigert carefully assembled the contraption. Thankfully, it worked! The Apollo 13 astronauts bravely persevered as their frigid "lifeboat" carried them around the Moon and back.

Had the accident occurred after Aquarius touched down at Fra Mauro, Swigert would have died in lunar orbit, and Lovell and Haise would have been stranded on the Moon. The crew of *Unlucky 13,* as their flight was called by some writers, was very lucky, indeed. Overcoming one problem after another, the crew safely returned to Earth on April 17, 1970. Without the skill and courage of Lovell, Haise, and Swigert, as well as the ingenuity and dedication of the engineers, technicians, and flight controllers on Earth, three men would have been lost in space. Apollo 13 was truly a space odyssey.

While none of its primary objectives were accomplished, one significant secondary goal was achieved by Apollo 13. The S-IVB third stage slammed into the Ocean of

Alan Shepard was America's first man in space, the only Mercury astronaut to reach the Moon, the oldest person at 47 years of age to leave footprints in the lunar soil, and the only golfer to tee off in one-sixth g! He commanded Apollo 14 which fulfilled Apollo 13's aborted mission. (NASA)

Storms due south of the crater Lansberg and 85 miles (134 km) west of the Apollo 12 seismometer. The impact was recorded as a seismic signal 20 to 30 times larger, and lasting 4 times longer, than the one produced by the Apollo 12 LM ascent stage. Two other ALSEP instruments detected a tenuous cloud of gas rising above the Moon's surface from the residual propellants in the S–IVB's disintegrated fuel tanks.

In the end, Apollo 13 returned scant knowledge about the nature of the Moon. Yet, it did provide an inspiring and lasting example of *human* nature at its finest.

21 Mission To Fra Mauro

The cause of Apollo 13's near-disaster was soon traced to what a NASA accident investigation board described as an unusual combination of human error and oversight, together with "a somewhat deficient and unforgiving design" — the same lethal combination that would destroy the Space Shuttle, *Challenger,* in January, 1986. Several weeks before the launch of Apollo 13, one of two oxygen tanks mounted in the Service Module (SM) had been damaged during a ground test. For a variety of reasons, it went unnoticed. As Lovell, Swigert, and Haise headed toward the Moon, the defective Number 2 tank exploded, puncturing tank Number 1 and the spacecraft's electrical power system. To prevent a similar accident in the future, the design of the SM oxygen tanks was modified, an extra tank isolated from the other two was installed, and an auxiliary battery was supplied for emergency use.

Having resolved the problem that crippled Apollo 13, NASA planners resumed their preparations for Apollo 14. Originally bound for a landing near the crater Littrow

Alan Shepard poses with the U.S. flag at Fra Mauro, the site Jim Lovell and Fred Haise had hoped to visit. The shadows (left to right) are from Antares, *astronaut Ed Mitchell, and an umbrella-like S-band antenna used on Apollo 12 and 14 to improve communications with Mission Control.* (NASA)

Alan Shepard removes tools from the "lunar rickshaw," a two-wheeled cart, designed to help the astronauts carry equipment and samples. Unfortunately, its tires frequently became mired in the lunar dust. (NASA)

on the southeastern shore of the Sea of Serenity, Apollo 14 was re-targeted to Fra Mauro, Apollo 13's destination. This was a site geologist Don Wilhelms called "the most important single point reached by astronauts on the Moon." Fra Mauro was significant because it offered scientists a chance to learn more about the Moon's early history.

Much of Earth's early history, recorded in its original crust, was obliterated eons ago through a variety of geological processes. Plate tectonics, for example, has played a major role in shaping Earth's surface. The movement of large rigid crustal plates around the globe is still occurring today, though at a snail's pace. In some places, the crust is being ingested into the hotter mantle below. In others, new crust is being formed (in the mid-oceanic ridges, for example). Because Earth is a geologically active, evolving planet, most pieces of its ancient past have long since disappeared.

The preliminary results of the first Apollo landings had suggested that the Moon had been a geologically inactive — dead — world throughout most of its long existence. Scientists surmised that pieces of its early crust might still be found, enabling them to more accurately determine the Moon's origins, age, and early history. Shortly before the flight of Apollo 14, geologist Robin Brett said, "By visiting Fra Mauro we hope to

In a pose familiar to terrestrial explorers, Ed Mitchell consults a map as he and Alan Shepard made their way toward Cone Crater. The rolling terrain and absence of familiar objects to serve as points of reference made it difficult to judge distances at Fra Mauro. (NASA)

sample the very bedrock of the Moon, material very different from that collected so far—material perhaps dating back to the very beginning of the Solar System… All in all, the Fra Mauro material should contain a great deal of new information about the early history of the Moon, and thus help us to better understand the formation of our own Earth."

Scientists believed that pieces of the Moon's early crust were blown to Fra Mauro by the giant impact that created the Imbrium Basin, but that material had long since been buried under the lunar regolith. So how could astronauts reach rocks that were 50 feet

(15 m) or more below the surface? The answer sat at the rim of Cone Crater, located roughly 3,600 feet (1,100 m) east of the primary landing site. "A large crater acts in many respects like a drill," Dr. Brett observed, "throwing out material from deep beneath the surface." The meteorite that punched out the 1,200 foot (370 m) diameter Cone Crater penetrated the local regolith and threw out pieces of Fra Mauro bedrock along its rim.

With this goal in mind, Apollo 14 was launched on January 31, 1971. Commander Alan Shepard, who had been America's first man in space, would become the oldest person, at age 47, to walk on the Moon and the only member of the original Mercury Seven astronauts to ever get there. The mission was a personal triumph for Shepard, who had been grounded for several years with an inner ear disorder. Cured after surgery in 1969, his dream of going to the Moon would at last be fulfilled. His crew included Command Module Pilot Stuart Roosa and Lunar Module Pilot Edgar Mitchell.

At several points during the mission, the success of Apollo 14 seemed in doubt, beginning with Roosa's first attempt to dock the Command Module (CM) *Kitty Hawk* with the Lunar Module (LM) *Antares*. In a normal docking, a probe on the nose of the CM slipped into a drogue mounted above the LM's upper hatch. Small capture latches on the tip of the probe engaged the drogue, after which a tunnel ring atop the LM's ascent stage meshed with the CM's latch assemblies, resulting in a hard dock and an airtight seal between the two spacecraft. By opening the CM's upper hatch, removing the probe and drogue, and opening the LM hatch, a tunnel was formed, permitting the crew to float between vehicles. Five times Roosa tried docking with *Antares,* and five times, he failed to get a hard dock. Meanwhile, Apollo 14 was on its way to the Moon. If Roosa were unable to extract the LM from the S-IVB, the lander would slam into the lunar surface with the third stage and the astronauts would be forced to circle the Moon and return to Earth without making a lunar landing. Finally, on the sixth try, *Kitty Hawk* and *Antares* were joined together.

Apollo 14 went into lunar orbit on February 4. Later that same day, Shepard and Mitchell powered up *Antares* and separated from Roosa in the CM. At this point, another problem arose. A faulty switch was sending an erroneous abort signal to the LM's guidance computer. This meant that as soon as the descent engine was fired, the LM's computer would automatically cancel the landing and head back toward *Kitty Hawk.* Engineers and technicians at the Massachusetts Institute of Technology, where the computer was designed, hastily devised a solution to the problem. Shepard assumed manual control of the engine throttle while Mitchell quickly entered a long series of numbers into the computer's keyboard. Another obstacle had been overcome.

As *Antares* began its descent to the Moon, its landing radar failed to get a lock on the lunar surface. The lock should have been achieved at about 30,000 feet (9,000 m). As the LM dropped lower and lower, Mission Control advised Mitchell to keep switching a circuit breaker off and on. At 22,713 feet (6,923 m) the radar locked on target, and Mitchell radioed, "Whew! That was close." Despite these problems, Shepard made the most accurate lunar landing of all, coming within yards (meters) of the planned landing site. Touchdown occurred at 4:17 A.M. EST on February 5, 1971.

Some 5½ hours later, Shepard became the fifth man to set foot on the Moon. Mitchell soon joined him, and the astronauts busied themselves positioning a color television camera, unfurling an S-band antenna, raising another American flag, and setting up an Apollo Lunar Surface Experiments Package (ALSEP) station some 600 feet (180 m) from the LM. To help them carry their equipment and samples, they had a two-wheeled lunar "golf cart" called a Modularized Equipment Transporter (MET). Designed to be pulled by hand, the MET's tires frequently became mired in the deep lunar dust. The rickshaw, as it was also known, was often easier to carry than to pull.

The main focus of the first Moonwalk was the deployment of the ALSEP station. A Swiss Flag and a Laser Ranging Retro-Reflector were also set up, as was a set of geophones for the Active Seismic Experiment (ASE). Mitchell used a thumper device to generate a series of small concussions on the lunar surface. These were sensed by the geophones and helped to determine the characteristics of the local regolith. Of the 21 charges in the thumper, only 13 fired, and only nine of these were detected by the geophones.

The astronauts also collected a number of rock and fine samples, including two rocks described as "football-sized," weighing about 5½ pounds (2.5 kg) each. These rocks were not actually the size of footballs, but they were relatively large. After nearly five hours, Mitchell and Shepard completed their initial traverse, bringing 44 pounds (20 kg) of lunar material into the LM with them. The tired Moonwalkers needed food and rest before their second Moonwalk.

Early the next morning, Shepard and Mitchell were on the surface of the Moon once again. They had planned to walk to the rim of Cone Crater, but the going was tough and they had a difficult time finding the rim among the boulder fields surrounding the crater. Finally, with oxygen and time running out, they were forced to turn back, not knowing that they had come within a few yards (meters) of their elusive goal. However, they at least managed to collect rock samples from very near the rim, returning a total of 96 pounds (43.5 kg) of lunar material to Earth.

Before blasting off to *Kitty Hawk* later that afternoon, Shepard hit the first golf balls on the Moon. That bit of fun did not impress the astrogeologists back home who were disappointed by the lack of time available at Cone Crater. They would have liked more samples to study, although the astronauts had obtained most of that for which they had come0 to Fra Mauro. The second Moonwalk had lasted just over 4½ hours.

Splashdown came on February 9, bringing Apollo 14 to a successful conclusion and putting the Moon program back on track.

22 Moon Buggy

Al Shepard and Ed Mitchell had had precious little time at the rim of Cone Crater. Their Portable Life Support System (PLSS) backpacks contained limited supplies of oxygen and cooling water, and communications with Earth depended on their maintaining a line of sight with the Lunar Module (LM), which relayed their transmissions. Shepard and Mitchell also had to carry or pull all of their equipment and samples by hand. The astronauts were therefore unable to carry back as many rocks and fines as the scientists desired.

The rocks which were collected seemed to indicate that the Imbrian impact had occurred about 3.8 billion years ago. Some of the rocks were even older; at 3.9 billion years, they predated the impact and represented material from the Moon's more distant past. Other rocks dated the formation of Cone Crater, a mere 25 million years ago. Most of the rocks were breccias, formed when angular fragments from older rocks were broken, melted, and cemented together by powerful meteorite impacts. All these findings supported the theory that the Moon's surface was shaped largely by meteoric rather than volcanic processes.

Unfortunately, the constraints placed on the astronauts' time and mobility made it very difficult to more fully explore the lunar surface and limited sample collections. New developments were being readied for the final Apollo expeditions to the Moon. The first three trips were called H missions. The last three were J missions. These flights incorporated modifications that would double the astronauts' time on the lunar surface.

The J missions also included an entirely new machine, a Lunar Roving Vehicle (LRV) which gave the astronauts a small "car" on the Moon. This greatly extended their range, and would overcome the problems encountered by Shepard and Mitchell during their long traverse to Cone Crater.

The Rover resembled a lunar dune buggy, leading some to call it a Moon Buggy. This nickname belied the fact that the Rover was a very sophisticated vehicle. It had to fit within a small compartment of the LM—which meant that it had to be folded up into a very small package for its trip to the Moon. After landing, the Rover was unfolded and deployed on the surface with minimum assistance from the astronauts. Just as important as the size constraints were weight limitations. The Rover weighed just 460 pounds (210 kg), yet would carry more than twice that weight, including two astronauts and their Portable Life Support Systems, communications gear, a television camera, scientific and photographic equipment, and 60 pounds (27 kg) of lunar samples.

Built by the Boeing Company, the Lunar Rover was powered by two 36-volt batteries. Each of the LRV's wire mesh wheels had its own quarter-horsepower electric motor, giving the Rover four-wheel drive and a top speed of around 8 mph (13 kph). The front and rear wheels had separate steering systems; if one failed, the vehicle could be operated

The first Lunar Roving Vehicle to reach the Moon is seen in this photograph taken near the end of the last Apollo 15 Moonwalk. The Rover greatly enhanced the astronauts' ability to explore the lunar surface. (NASA)

with the other. The Rover could climb and descend slopes as steep as 25 degrees, and negotiate small obstacles and crevices.

The Rover was 10 feet 2 inches (310 cm) long, 6 feet (182.9 cm) wide, and had a wheelbase of 90 inches (225 cm). It could travel a cumulative distance of 40 miles (65 km), although a 6 mile (9.6 km) radius from the landing site was established as a safety measure so that the astronauts could walk back to the LM in case the Rover broke down. Still, this gave the astronauts 113 square miles (293 sq km) to explore, or 10 times the area that could be reached on foot.

Seated side-by-side, the astronauts drove the Rover with a T-shaped hand controller that was mounted between them. This "stick" controlled steering, speed, direction (forward or reverse), and braking. Just in front of the stick was a small control and display

A broken fender on the Apollo 17 Moon Rover was repaired using clamps and a spare lunar map. Without it, the Rover's wire mesh wheels kicked up lunar dust — showering the crew and equipment. Note the chevron-shaped titanium treads riveted to the zinc-coated piano wire tires. Jack Schmitt is seated in the Rover's right-hand seat with his back to the camera. (NASA)

console which provided information about the vehicle's performance. Navigational data was also given, including the direction and distance back to the LM. A suitcase-sized Lunar Communications Relay Unit (LCRU), mounted at the front end of the Rover, maintained direct communications with Earth. When the Rover was stopped, a High-Gain S-band antenna enabled Mission Control to monitor the astronauts' activities via a remotely-controlled television camera. The rear of the Rover had room for a tool carrier. Equipment and samples could be stowed under the Rover's seats.

The Lunar Roving Vehicle was an amazing little machine that greatly enhanced the astronauts' ability to explore the Moon. It proved invaluable during each of the final Apollo missions.

23 Hadley Base

If Apollo 11 had accomplished what we did on Apollo 15, the American public would not have lost their interest in Moon missions, and Apollo 18, 19, and 20 would never have been cancelled.   Private observation of a senior NASA public affairs officer, 1971

The statement may be somewhat debatable, but there is no doubt that the flight of Apollo 15 was, as mission Commander David Scott put it, "exploration at its greatest!" Yet this first Apollo J mission did not generate the sort of public interest that accompanied Apollo 8's first trip around the Moon, Apollo 11's first lunar landing, or the dramatic flight of Apollo 13.

Television ratings and public opinion aside, Apollo 15 was one of history's greatest voyages of discovery. For those who followed the flight on television, it was almost like watching a science fiction film, as viewers watched Scott and Lunar Module Pilot James Irwin walking on the Moon against a backdrop of lunar mountains, craters, boulder fields, and a spectacular, V-shaped gorge.

The area where Mare Imbrium abuts the Apennine mountain range is known as the Apennine Front. Located on the southeastern edge of Mare Imbrium, in an area called Palus Putredinis (Marsh of Decay), the Hadley-Appenine landing site was the northern-most spot reached by the Apollo missions. It was chosen for Apollo 15 because it offered a great variety of geological features, including the soaring Mount Hadley, which reaches nearly 15,000 feet (4,600 m) into the lunar sky, and Hadley Rille, which drops 1,150 feet (350 m) below the ground level. Apollo 15 landed on a mare plain within reach of the rille and another tall mountain, Hadley Delta, named for John Hadley, an 18th Century British scientist.

Geologist-Astronaut Harrison "Jack" Schmitt, who would later fly on Apollo 17, explained the significance of the Hadley-Apennine site at a press briefing shortly before Apollo 15 reached the Moon:

The Imbrium Basin formed very early in the geologic history of the Moon. This, we have deduced from the analysis of the Apollo 14 rocks and from the general geologic character of the Imbrium Basin relative to the other features on the Moon. The Apennine Front represents one of the upthrown rims, one of several rings of mountains, that were created by [the] Imbrium impact event. By being upthrown — that is, the outer part thrown up and the inner part of the basin dropped down — it exposed a section of lunar crust...some 15,000 feet [4,600 m] of vertical rock section.

Thus, samples from the Apennine Front might include pieces of the pre-Imbrian lunar crust. Likewise, the one mile (1.6 km) wide, 100 mile (160 km) long Hadley Rille cut through the mare surface, likely exposing different layers that represented successive lava flows. These features resemble the layering observed in many Earth mountains and

This photograph of Jim Irwin, the Apollo 15 Lunar Rover Vehicle, and Hadley Rille was taken by Dave Scott from St. George Crater at the foot of Hadley Delta. The floor of the deep rille was found to be filled with broken rocks from the multiple layers of basalt on the upper walls, evidence of successive lava flows across the mare. (NASA)

canyons, and although formed differently, the lunar layers promised to expose a similar record of geologic history.

The third member of Apollo 15, Command Module Pilot Alfred Worden, was to study the Moon's surface from lunar orbit with eight experiments mounted in a previously empty section of the Service Module called the SIM (Scientific Instrument Module) bay (for more about the SIM bay, see Chapter 24). Thus Apollo 15 would be the most ambitious Moon mission so far.

Lift-off came at 9:34 A.M. EDT on July 26, 1971. The all-Air Force crew dubbed their Lunar Module (LM) *Falcon,* for the official USAF mascot, and their command ship *Endeavour,* for the famous scientific sailing ship commanded by the eighteenth century English explorer, Captain James Cook. The three men reached lunar orbit three days later, and Scott and Irwin descended to the surface the following evening, July 30th. At 6:15 P.M. EDT Scott reported, "Okay, Houston. The *Falcon* is on the plain at Hadley."

Two hours after touchdown, Scott opened the top hatch of the LM and stuck his head out for a look around Hadley Base. "All of the features around here are very

Dave Scott takes a documentary photograph of a lunar sample on the flank of Hadley Delta. The object at the right is a "gnomon," used to indicate the local vertical and to calibrate photographs. In the background is the Lunar Roving Vehicle, its High-Gain antenna pointed toward Earth, and the Apennine Mountains over which Falcon *flew during its descent to the Moon's surface. (NASA)*

smooth. The tops of the mountains are rounded off. There are no sharp, jagged peaks or large boulders apparent anywhere." This description was in sharp contrast to the stark lunar scenes that space artists and science fiction film makers had depicted for so many years.

The next morning, Scott and Irwin climbed down *Falcon's* ladder to the lunar surface. Almost immediately, they unpacked and test-drove their Rover. The vehicle performed beautifully and the astronauts drove to St. George Crater at the base of Mount Hadley Delta, where they had an awe-inspiring view of Hadley Rille.

Studying a nearby rock, Jim Irwin remarked, "That looks fairly recent, doesn't it, Dave?"

Back in Houston, Scientist-Astronaut Joe Allen jumped into the conversation with, "It probably *is* fresh — probably not older than three and a half billion years."

To which Scott replied, "Can you imagine that, Joe? Here sits this rock, and it's been here since before creatures roamed the sea on our little Earth."

After collecting six bags of rocks, four bags of soil, and two double core tube samples, the astronauts headed back to the LM to deploy their Apollo Lunar Surface Experiments Package (ALSEP). Drilling two 10 foot (3.0 m) holes for a new lunar Heat Flow Experiment proved to be a very difficult task, even with the help of a battery-powered Lunar Surface Drill, and was temporarily abandoned at the end of their first Moonwalk. The purpose of the experiment was to measure the amount of heat flowing from the Moon's interior, heat generated by the decay of radioactive elements such as thorium, uranium, and potassium. Sensors lowered into the holes would measure temperatures and help to establish the thermal history of the Moon.

This is Hadley Base looking south toward St. George Crater. Jim Irwin is behind the Rover, and Hadley Delta is behind Falcon. The LM is slightly tilted as Dave Scott had landed on the rim of a shallow crater. (NASA)

Before climbing back into the LM, Scott and Irwin set up a laser reflector and solar wind experiment (Swiss Flag). They had spent an exhilarating, though tiring, 6½ hours exploring Hadley.

Early the next morning Scott and Irwin began their second Moonwalk. This time, they drove the Rover back to the Apennine Front at the base of Mount Hadley Delta. Knowing that the ejecta along the rim of a crater represents the deepest material blasted to the surface by a meteorite impact, they carefully surveyed the rocks around the rim of a small crater.

"Oh, man. Look at that!" Jim Irwin exclaimed.

"Guess what we just found," Scott answered. "I think we found what we came for."

"Crystalline rock, huh?"

"Yes, sir," Scott enthusiastically confirmed. "I think we might have ourselves something close to anorthosite." (Anorthosite rocks are composed primarily of the mineral, plagioclase, which scientists think was the main constituent of the primordial lunar crust.)

The press quickly dubbed the astronauts' discovery the "Genesis" rock, and analysis later validated that nickname. Sample number 15415 proved to be 4.5 billion years old, nearly as old as the Moon itself!

Before closing out their second Moonwalk, the astronauts finished drilling the second hole for the Heat Flow Experiment and erected an American flag at Hadley Base. Another well-deserved night's rest prepared them for a final Moonwalk early the next morning.

During their third, and last day, on the Moon, Scott and Irwin explored the east rim of Hadley Rille. Across the rille, they could see several well-defined layers of basalt in the west wall—evidence that successive lava flows had, indeed, covered the mare. Bedrock samples from the edge of the rille (the only place on the Moon where intact bedrock was ever collected) turned out to be 3.3 billion years old.

Before rejoining Worden in *Endeavour* later that same afternoon, Scott and Irwin parked their Rover some distance behind the LM so that the vehicle's camera could televise their lift-off—another first for Apollo 15. The Moon buggy had carried the explorers 17.5 miles (27.9 km) over the lunar surface, and had made it possible for the men to bring back 170 pounds (77 kg) of rocks, fines, and core samples that spanned more than a billion years of the Moon's earliest history. The ALSEP station left behind on the Moon would continue to send data, as would a small scientific subsatellite that was launched into lunar orbit from *Endeavour's* Scientific Instrument Module bay shortly before the Service Propulsion System engine was fired on August 4.

During the homeward journey, Worden made a spectacular spacewalk some 197,000 miles (317,000 km) above Earth to retrieve film magazines from the panoramic and mapping cameras mounted in the SIM bay.

Even though one of the Command Module's three parachutes collapsed seconds before the Apollo 15 astronauts splashed down safely in the Pacific on August 7, 1971, the mission was a stunning success. It marked a true milestone in lunar exploration. The crew brought home a wealth of scientific data and a treasure-trove of lunar samples, including a piece of the Moon's original, ancient crust. Three days on the lunar surface, three record-breaking Moonwalks, the first car on the Moon, a remotely-controlled color television camera, and a breathtaking landing site surrounded by the towering Apennine mountain range and a meandering canyon, Hadley Rille, had all come together to provide the most thrilling Moon flight yet.

24 Apollo Orbital Science

The last three Apollo J missions used improved Lunar Modules and had additional oxygen and power supplies that extended the flight duration of the command ship. To make good use of the extra time in lunar orbit, a new Scientific Instrument Module was installed in a previously empty section of the Service Module (SM) called the Scientific Instrument Module (SIM) bay.

About 4½ hours prior to lunar orbit insertion, the experiments in the SIM bay were exposed to the space environment when a panel on the side of the SM was jettisoned. Once in orbit around the Moon, the experiments were controlled by the Command Module (CM) pilot. One of his primary objectives was to obtain high-quality photographs of the lunar surface, using two sophisticated cameras—one for mapping, and the other for scientific studies. To accomplish these goals, the attitude of the Command/ Service Module (CSM) with reference to the stars and its exact position above the lunar surface had to be determined with great accuracy. At the same time that the Moon was being photographed, a stellar camera pointed perpendicular to the main cameras took pictures of star fields, while a laser altimeter precisely measured the distance to the lunar surface.

A mapping camera system, built by Fairchild Camera and Instrument Corporation, used a 76 mm cartographic lens to eliminate distortions. The camera produced medium-resolution images, on which, distances between points could be precisely measured for the purpose of compiling medium-scale topographic maps.

The 610 mm panoramic camera, manufactured by Itek Corporation, was similar to the reconnaissance systems in use at the time by the U.S. Air Force. The camera produced high-resolution images, revealing objects on the lunar surface as small as a desk.

These pictures would be useful in studying the Moon's geology.

To obtain good results, the two cameras had to compensate for the spacecraft's velocity and height, much like the cameras on Lunar Orbiter. However, the Apollo imaging systems offered far greater detail than those of Lunar Orbiter. The major drawback of the Apollo imaging systems was that they covered a relatively small area beneath the spacecraft's flight path and photography was therefore confined to the Moon's equatorial regions. However, overlapping images did provide outstanding stereoscopic views of countless lunar features.

The large-format film for these cameras was stored in two removable cassettes. After leaving lunar orbit, the CM was depressurized, and its main hatch opened. The CM pilot then made his way to the SIM bay using a series of hand rails and a pair of "golden

The Apollo 15 Commmand/Service Module as seen from the Lunar Module in lunar orbit. The open Service Module Scientific Instrument Module (SIM) bay can be seen with its array of mapping cameras and experiments. (NASA)

Apollo 17 Command Module Pilot Ron Evans retrieves the mapping camera film cassette (the cylindrical object near his elbow) from the Service Module Scientific Instrument Module (SIM) bay during the journey back to Earth. His spacewalk lasted 1 hour and 7 minutes. Lunar Module Pilot Jack Schmitt took this photograph from the Command Module's open hatch. (NASA)

slippers" (foot restraints). He retrieved the cassettes, which then rode back to Earth inside the CM.

In addition to its camera systems, the SIM bay housed several scientific instruments. Three of them, the Chemical Group, were designed to study the Moon's geochemical properties:

- The **Gamma Ray Spectrometer,** mounted on a 25 foot (7.6 m) extendible boom, measured the chemical composition of the lunar surface directly beneath the spacecraft.

- An **Alpha Particle Spectrometer** (APS) located radioactive elements buried below the surface of the Moon. Decaying thorium and uranium produce radon gas which, in turn, produces alpha particles which escape into space. By measuring this alpha radiation, the APS was able to indirectly locate these two radioactive elements.

- The **X-Ray Fluorescence Spectrometer** detected fluorescent x-ray patterns generated by solar x-rays striking the lunar surface and exciting the atoms of lunar material. Since different elements generate different X-ray patterns, the rock types in various regions of the Moon could be identified and quantified.

Because an atmosphere would have prevented the particles from being detected in lunar orbit, the near-vacuum environment of the Moon was crucial for these instruments to make their measurements. Another instrument, a Mass Spectrometer, located on a second extendible boom near the rear of the spacecraft, studied the composition and density of gas molecules along the CSM flight path. Some of the gases it detected were from the solar wind. Others were vented by the CSM and Lunar Module. But a few were possibly produced by the Moon itself. The Mass Spectrometer was designed to measure this tenuous lunar atmosphere.

Just before the astronauts left lunar orbit, a small Particles and Fields Subsatellite was ejected from the SIM bay. Powered by solar cells, the subsatellite detected solar wind particles and measured the Moon's weak magnetic field (less than 1/1000th of Earth). The subsatellite was also used to study variations in the Moon's gravity field, refining the mascon data obtained previously by Lunar Orbiter. Subsatellites were ejected on both the Apollo 15 and 16 missions.

Apollo 17's SIM bay carried several new instruments and deleted others, although the mapping and panoramic cameras were an important part of its payload. When his crewmates were on the lunar surface, the CM pilot also made countless observations of the Moon with two of nature's finest instruments—his own eyes. On Apollo 15, for example, Al Worden spotted what appeared to be small, dark, cinder cones (dead volcanic craters) near the crater Littrow on the southeastern edge of Mare Serenitatis. Based in part on Worden's visual observations, Apollo 17 was targeted for that area of the Moon a year and a half later.

25 Exploring the Lunar Highlands

With the flight of Apollo 15, lunar astronauts had collected samples from four sites on the Moon, most of them representing mare plains. But only 17 percent of the Moon's total surface area (and just 2.5 percent of its far side) is covered by dark, basaltic maria. Therefore, as the Apollo program neared its conclusion, scientists were anxious to retrieve material from the lunar highlands—the brighter regions of the Moon as seen from Earth. Because the heavily-cratered highlands were more rugged than the smoother plains, they had been avoided for safety reasons during the early Apollo missions. With four landings safely behind them, NASA planners had gained confidence in the Lunar Module (LM) and in their own abilities. Now Apollo 16 was being targeted for loftier terrain near the crater Descartes.

While driving "home" from North Ray Crater at the end of the third Apollo 16 Moonwalk, Charlie Duke took this photograph of the Lunar Module Orion (near the center of the picture) and Stone Mountain (the ridge to the left). In the foreground are the Lunar Rover's High-Gain antenna, and the remotely-controlled television camera covered by a lens hood. (NASA)

Astronaut John Young walks away from the Apollo Lunar Surface Experiments Package (ALSEP) site at Descartes. In the background are the Lunar Tri-Axis Magnetometer, the Radioisotope Thermoelectric Generator, the Lunar Surface Drill, and a rack with bore stems for the drill. (NASA)

The area around Descartes had been repeatedly bombarded over the eons by meteorite impacts. Craters overlapped craters, and the degraded appearance of the older ones gave silent testimony to their relative ages. The Apollo 16 site was a bit different from the surrounding highlands, in that it encompassed two distinct types of terrain: the Cayley Plains and the Descartes Formation. Both appeared to be of volcanic origin, perhaps even younger than the maria. Geologists were particularly interested in studying this unique region in order to help them better understand the evolution of the Moon.

(One other landing site had been considered for Apollo 16. The crater Alphonsus, where Ranger 9 had crashed, had drawn the attention of scientists in 1958, when Soviet astronomer Nikolai Kozyrev reported seeing a reddish glow emanating from its central peak. Similar presumably gaseous emissions were seen by other astronomers near the crater Aristarchus in 1963. These sightings were referred to as "lunar transient events" and were the focus of much speculation—some scientists had cited them as evidence of a geologically active Moon. Subsequent lunar missions, starting with Ranger, had seemed to rule out that possibility, so mission planners settled on sending Apollo 16 to the Cayley Plains near the Moon's center, as viewed from Earth.)

Mission Commander John Young, who flew to the Moon on Apollo 10, would spend three days on its surface with Lunar Module Pilot Charles Duke. A Lunar Roving Vehicle would enable them to drive to the Descartes Formation so that two very different geological units could be explored. Meanwhile, Command Module Pilot Ken Mattingly would study the Moon from lunar orbit, the assignment from which he had been bumped for medical reasons on the ill-fated Apollo 13 mission. Now he would get his chance to see the Moon, close-up.

As with most previous flights, the Apollo 16 mission experienced some problems and surprises. The mission began with lift-off from the Kennedy Space Center at 12:54 P.M. EST on April 16, 1972. Shortly after translunar injection, the crew noticed that some of the silver paint used for thermal protection on their LM, *Orion*, was blistering and flaking off. Engineers decided, however, that this problem would not present a hazard. After *Orion* separated from the Command Module *Casper* for descent to Descartes on April 20, Mattingly reported a more ominous malfunction. When he tested the com-

Charlie Duke stands by the Lunar Rover at Station 4, halfway up the side of Stone Mountain. In the foreground are the tire tracks left by the Rover, whose right wheels bounded over the large rocks in the lower right-hand corner of the picture. (NASA)

mand ship's back-up steering system, *Casper's* main rocket engine vibrated back and forth. These unexpected oscillations did not occur while testing the primary steering system. However, since the SPS engine had to work perfectly to send the astronauts back home, any failure in either the primary or back-up steering systems would necessitate cancellation of the lunar landing and an early return to Earth.

For six hours, NASA engineers studied the problem by running simulations and assorted tests. In the end, they determined that the oscillations did not pose a serious threat to the engine's performance. A "go" was given for lunar descent. Young described the long wait as "a real cliff-hanger." The delay changed the flight plan considerably, shortening the time spent in lunar orbit by an entire day. But Young and Duke would still have time to accomplish most of their tasks on the lunar surface. Touchdown on the Moon came at 9:23 P.M. EST.

The southernmost point reached by an Apollo mission, the Apollo 16 site, sat on a plateau 8,000 feet (2,400 m) above the level of Tranquillity Base, 150 miles (240 km) to the northeast. Owing to their late arrival at Descartes, the astronauts put off their first Moonwalk until the next morning, when they planned to unload their Rover, raise the flag, and deploy another Apollo Lunar Surface Experiments Package (ALSEP). They were also scheduled to set up the first astronomical observatory on the Moon, a Far Ultraviolet Camera-Spectrograph which had been designed to study interstellar hydrogen gas and related phenomena, visible only in the far ultraviolet band of the electromagnetic spectrum. Such observations are impossible from Earth because our atmosphere filters out far ultraviolet light.

At 11:58 P.M. EST on April 21, 1972, Young became the ninth man to set foot on the Moon, declaring as he did so, "There you are, our mysterious and unknown Descartes Highland plains. Apollo 16 is going to change your image." Young's words were prophetic. After Duke joined Young, the astronauts started collecting their first lunar samples. They were all breccias, those coarse-grained rocks composed of angular fragments fused together by meteorites striking pre-existing rocks.

Speaking from Earth, Capsule Communicator Tony England asked the astronauts, "Have you seen any rocks that you're certain aren't breccias?"

"Negative," Duke answered. "I haven't seen any that I'm convinced is not a breccia."

The volcanic rocks that scientists had expected to see at Descartes were nowhere to be found. The reason was quite simple: none were there. This surprise forced the scientists to revise their thinking about the region's past. Clearly Descartes was the product of meteorite bombardment rather than highland volcanism. Surprises such as this demonstrate the true value of exploration — discovering the unexpected is every bit as important as proving preconceived theories.

Unfortunately, the first Moonwalk resulted in a disappointing setback. While deploying the ALSEP station, Young inadvertently tripped over an electrical cable connected to the Heat Flow Experiment. The cable snapped and the experiment was lost. But despite this loss, the Apollo 16 ALSEP was an important addition to the growing network of automatic scientific stations deployed on the lunar surface. After more than seven hours on the Moon, including a short drive in their Lunar Rover, the astronauts climbed back into *Orion* for dinner and a night's rest.

The next day they were out exploring again, this time driving south to Stone Mountain and the Descartes Formation. Stone Mountain was not nearly as impressive or as high as Mount Hadley and the Apennines, but after driving the Rover up Stone Mountain's 20 degree slope, Young and Duke reached a spot halfway up the 1,600 foot (490 m) mountain—the highest elevation reached by any Apollo astronauts. "What a view!" Duke exclaimed as he looked around at the rolling Moonscape.

During their excursion Young and Duke retrieved samples from a bright, fresh crater, South Ray. The astronauts had to settle for ejecta tossed some distance from the crater, as large blocks immediately surrounding its white rim made the area impossible to traverse with the Rover. Following this second, seven hour Moonwalk, the astronauts returned to *Orion* for the night.

On the third day, Young and Duke headed north three miles (4.8 km) to the rim of a 3,000-foot (914 m) diameter crater they dubbed North Ray. The astronauts were careful not to get too close to the crater's rim because, as Young later explained, "If we had fallen in, we would not have been able to get out."

Near the rim, they chipped off samples from a giant-sized boulder, House Rock. This last Moonwalk was a short one, less than six hours, because with the revised schedule the astronauts had to return to *Orion* for lunar lift-off. Along the way they chalked up a speed record with their Rover: 13.5 mph (22 kph). They loaded up their 213 pound (96.6 kg) haul of lunar samples, closed *Orion's* hatch, and blasted into orbit.

The next morning, April 24, *Casper's* main engine fired without a hitch and sent the astronauts on their way home. Mattingly enjoyed a spacewalk between Earth and the Moon to retrieve film canisters in the Scientific Instrument Module (SIM) Bay. Splashdown in the Pacific came at 2:45 P.M. EST on April 27, 1972.

When the rocks were unloaded at the Lunar Receiving Laboratory in Houston, the astronauts' observations were confirmed: space debris, not volcanism, was responsible for creating the Descartes Highlands. The Cayley Plains had been formed by molten ejecta splashed out by the Imbrium impact 1,000 miles (1,600 km) to the northwest. The older, rougher Descartes Formation was laid down by the impact that created the Nectaris Basin 375 miles (600 km) to the east. Some material might even have come from the Orientale Basin 2,200 miles (3,500 km) to the west.

Conclusion: the ancient lunar highlands are the battered remnants of the Moon's original crust.

26 Tools for the Moon

During their return voyage from Descartes, the Apollo 16 astronauts held a press conference. One question dealt with chipping off samples from large, lunar rocks using a geological hammer.

"It's pretty hard to do in a pressure suit," John Young observed.

Why? Because the astronauts were essentially working inside a balloon. Every time they wanted to move an arm, leg, or hand, they had to overcome the resistance created by the inflated spacesuits. Since much of their work involved gripping rocks, handles, and tools, most of the Apollo astronauts were plagued by sore fingers. Special tools were developed to make their tasks easier. For the last three Apollo missions, the lunar spacesuits were also modified with the addition of a waist-joint that improved mobility, enabling the explorers to bend while gathering samples, or to sit while riding on the Rover.

Scientist-Astronaut Harrison "Jack" Schmitt helped develop the tools used on the Moon, a job he was qualified to tackle given his doctorate in geology. Schmitt also played a key role in training all the Apollo crews for their various activities on the lunar surface. (Later, Schmitt had his own chance to work on the Moon during the Apollo 17 mission.)

Since the focus of all six lunar landing crews was geology, the astronauts carried the basic tools used by all geologists in the field, "the field" being any place where rocks can be observed and collected *in situ*, that is, in their natural setting and position.

The most basic geological tool is the geological hammer. The lunar version served three functions: it was a sampling hammer to chip off small samples from larger rocks, a pick or mattock (digging tool), and a hammer to drive lunar core tubes. The head had a small hammer face on one end, a broad, horizontal blade on the other, and large, hammering flats on the sides. A long handle made the Lunar Hammer easier to grip while wearing bulky lunar gloves. The lower end of the handle had a quick-disconnect feature that allowed the hammer to be attached to an extension handle for use as a hoe.

The extension handle was an aluminum tube, roughly one yard (1 m) long, which could be attached to other tools so that the astronauts would not have to bend or kneel to use them. For example, the adjustable sampling scoop, similar to a garden scoop, could be attached to the extension handle for gathering soil samples. One astronaut scooped up the sample while his partner held a collection bag. Similarly, a lunar rake attached to an extension handle enabled the astronauts to collect small rocks buried just beneath the surface. The soil fell through the stainless steel tines while the rocks caught in the rake's basket. Another tool, lunar tongs, was developed to help the astronauts grab small rocks from the surface while standing. Other tools, such as a brush for wiping off dust, were also used. Most of the astronauts' tools were stowed in a special rack mounted at the rear of the Lunar Rover, while a few were carried on their backpacks.

Apollo 17 Geologist-Astronaut Jack Schmitt uses a special lunar rake to collect coin-sized rock samples from the lunar topsoil. (NASA)

Core tubes were designed to be hammered or augered into the lunar soil in order to acquire samples of vertical layering in the upper regolith. For instance, a nearby meteorite strike might throw a fresh layer of soil on top of an older layer. A million years later, another meteorite hit might cover that layer, then another, and another, each layer recording the recent history, relatively speaking, of a given site. The short, hand-driven core tubes were supplemented on the J missions by a 10 foot (3 m) model used with the Lunar Surface Drill. The battery-powered drill was also used to make the holes needed for the Lunar Heat Flow Experiment (see Chapter 19).

Lunar documented sample bags resembled today's plastic sandwich bags. They came in 20-bag dispensers, each one with a number, so that scientists on Earth could match the rocks with the astronauts' descriptions of where they came from on the Moon. For

Astronaut Jim Irwin digs a small trench, using a scoop mounted on an extension handle during the Apollo 15 mission. Part of Mount Hadley can be seen behind him in this view looking to the north. (NASA)

example, on Apollo 15, scientists knew exactly where to find Dave Scott's "Genesis" rock when the rock boxes were unpacked. Astronauts also took several photographs of each documented sample, showing its exact placement on the Moon before the rock was collected. Then they photographed the scene from whence the sample was removed. To interpret the significance of each sample returned from the Moon, it was critical to document as much information as possible regarding its setting. With rocks being tossed about, modified, and transplanted by meteorite impacts over the eons, an individual sample's relationship to other rocks and nearby craters could offer clues to its past history. Earthbound geologists practice many of these same field techniques and procedures.

To help them in these tasks the astronauts used a gnomon, a 12 inch (0.3 m) rod freely mounted on a tripod. The gnomon indicated the local vertical, showed the position of the Sun, and served as a size scale, It also included black and white and color charts which allowed technicians on Earth to accurately determine contrast and hues when processing the film. The gnomon can be seen in many of the images taken on the lunar surface.

Because the LM's Ascent Stage had limited thrust, a limited amount of lunar samples could be carried back into lunar orbit and then home. Each crew therefore brought a small spring scale to weigh their treasures before loading them on board.

The Apollo Lunar Sample Return Containers—rock boxes—were suitcase-sized and made of aluminum. Most of the lunar samples were returned in two of these, but a Special Environmental Sample Container was also carried to preserve a carefully-selected rock in the lunar environment. Vacuum seals prevented any gases or contaminants from escaping or entering the container during the journey back to Earth.

The Apollo astronauts had precious little time to enjoy the view. Their tightly-controlled schedules focused on assorted work assignments and on getting the greatest scientific return from each minute of every mission.

27 The Valley of Taurus-Littrow

Apollo 17, the twentieth century's last manned voyage to the Moon, brought the curtain down on mankind's greatest adventure with a grand finale. It was a bittersweet finish for the people who had worked so long and so hard to make the Moon missions possible.

The last lunar mission began on December 7, 1972, at 12:33 A.M. EST with the spectacular nighttime lift-off of a Saturn V Moon Rocket. It ended with a string of records: the longest Apollo flight (12½ days); the longest time spent in lunar orbit (6 days); the longest time spent on the lunar surface (75 hours); the longest distance driven on the Moon (22 miles or 35 km); and the heaviest load of lunar samples returned to Earth (243 pounds or 110 kg). The flight also included some of the Apollo program's most impressive television images, and the first geologist to explore the Moon's surface, Dr. Harrison H. ("Jack") Schmitt.

The Apollo 17 crew was commanded by Apollo 10 veteran, Eugene Cernan. Ronald Evans served as Command Module Pilot and Jack Schmitt was the Lunar Module Pilot. The timing of their midnight launch, which lit up the sky over Cape Kennedy like the Sun, was made necessary by celestial mechanics; in order to reach its landing site near the southeastern rim of Mare Serenitatis under the desired lighting conditions, Apollo 17 had to leave Earth in the middle of the night. Those who witnessed the event will never forget the blinding light and thundering roar of the Moon rocket as it rose majestically from Pad 39A.

Mission planners had considered three sites for this final expedition to the Moon. One was the crater Alphonsus, which had also been considered for Apollo 16. The transient events (reddish glows) which had been seen there in the past still intrigued some scientists. "We really don't have a good explanation for transient events if, indeed, they do occur," Schmitt stated, during an interview in 1994. "There's one school of thought—I don't share it—that says when an astronomer or geologist stares at the Moon long enough, he's likely to see anything!" However, Alphonsus was eliminated as

"Comparable to the Grand Canyon in scale and grandeur," Jack Schmitt wrote after his mission, "the Valley of Taurus-Littrow extends some 20 miles through the ring of massifs surrounding the plains of the Serenitatis basin." In this orbital view taken from the Lunar Module, South Massif and the bright landslide material at its base are seen near the center of the picture. The Command/Service Module appears as a dark speck set against the massif. The Lunar Module Challenger *landed near the cluster of small craters visible on the valley floor.* (NASA)

During the first Apollo 17 Moonwalk, Gene Cernan took the Lunar Rover for a short test drive. South Massif towers over the Lunar Module Challenger *in the background.* (NASA)

a landing site for the same reasons invoked prior to the Apollo 16 mission, even though it was believed that interior gas likely escapes to the surface on occasion at the site.

The crater Gassendi offered another option. The large 68 mile (110 km) wide crater sits on the northern edge of Mare Humorum (Sea of Moisture) in the southwestern quadrant of the Moon's near side. It features a fractured floor and central mountain peaks, objects of interest to many scientists.

However, the Taurus-Littrow site was finally selected because it seemed to offer a chance to collect some young volcanic rocks, something that had not been found by any previous Apollo crew. The actual landing spot was in a beautiful valley surrounded by towering massifs (mountains) located south of the crater Littrow and west of the Taurus Mountains in the Moon's northeastern quadrant. Cernan and Schmitt planned to sample huge boulders near the base of one mountain, visit a landslide at the bottom of another, and cross a scarp (a line of cliffs, produced by faulting).

One of the main attractions at Taurus-Littrow was the valley's dark basaltic floor, which geologists thought might offer some evidence of relatively recent volcanic activity. This belief was supported by Al Worden's observations from lunar orbit during Apollo 15 of what appeared to be volcanic cinder cones. The interest in young material

During Apollo 17's third Moonwalk, Cernan and Schmitt drove the Lunar Rover to the base of North Massif, where they investigated a garage-size boulder that had rolled down the mountain and split into five pieces. Here Jack Schmitt takes samples in the vicinity of one of the pieces. (NASA)

was driven by the scientists' desire to know when most geological activity on the Moon had ceased. While previous flights had provided clues to understanding the Moon's more distant past, this last mission would hopefully uncover some clues regarding its more recent history—the last 3 billion years!

Cernan and Schmitt reached the Valley of Taurus-Littrow at 1:55 P.M. EST, on the afternoon of December 11, 1972. Five hours later, Cernan descended the ladder of the Lunar Module (LM), *Challenger* and dedicated "the first step of Apollo 17 to all those who made it possible."

Schmitt followed a few minutes later, and the astronauts raised the flag, unloaded the Lunar Rover, and set up the last Apollo Lunar Surface Experiments Package (ALSEP). During their first traverse, they inadvertently knocked off part of the Rover's right rear fender. As a result, they were showered by Moon dust as they drove. Later, while they enjoyed their first night's rest in *Challenger,* Mission Control came up with a fix for the damaged Moon buggy. At the start of their second Moonwalk, the astronauts clamped several unused maps onto the remaining fender. The repair worked perfectly.

It was a good thing that it did, because Cernan and Schmitt covered a lot of ground during their second day on the Moon, some 12 miles (19.5 km). They drove to South Massif and to the crater Nansen, which was half-buried by a landslide that appeared to have been triggered 100 million years before, when ejecta from the crater Tycho struck the top of the mountain. This is astounding when you consider that Tycho is 1,250 miles (2,000 km) southwest of Taurus-Littrow! But Tycho's far-reaching rays support this hypothesis.

On the way back to the LM, the astronauts made a remarkable discovery. Stopping at a small dark crater, Shorty, Schmitt suddenly exclaimed, "Oh, hey! Wait a minute... There is orange soil!" Schmitt stirred the soil with his feet. A few minutes later he remarked, "If there ever was something that looked like a fumarole alteration, this is it."

Schmitt's comment caught the attention of geologists back on Earth because a fumarole is a volcanic vent which emits gaseous vapor. Had a lunar volcano finally been found?

Jack Schmitt examines "Split Rock" during Apollo 17's third Moonwalk. This spectacular view looks south across the Valley of Taurus-Littrow from the base of North Massif. (NASA)

Dr. Schmitt gave the answer in an interview 22 years after the mission. "It [the soil] turned out to be made up of little beads of orange glass like we find around volcanic fire fountains in areas such as Hawaii. And the reason we were so excited was that in our pre-mission planning we had anticipated that this particular crater might be something different. It was very dark and superimposed on an avalanche deposit. And there was the faint possibility in our thinking — while we were still on Earth — that it might be a little volcano. So in the context of it being a volcano, we had speculated that we might find some 'alteration' as we call it — some changes of color and oxidation state around that crater."

Dr. Schmitt went on to explain, "When we got there it was obvious that it wasn't a volcano — it was just another impact crater. But still, the thought process had begun in my mind to be looking for something different. And sure enough, down there on the ground and on my boots was this orange material that turned out to be clearly unusual in the context of what we had seen on the Moon before."

Shorty Crater was not a volcano, but the orange beads embedded in its soil were created in an ancient lunar fire fountain. And their discovery was important. According to Dr. Schmitt. "The reason [the orange soil] turned out to be significant is that through various geochemical tests we were able to determine that it came from very deep within the Moon — probably as deep as 300 miles (500 km). And it represents the only sample we have of the deep interior of the Moon." The glass beads had formed some 3.7 billion years ago, during the period when dark, lava flows were covering the valley floor at Taurus-Littrow.

Most of the samples returned by Apollo 17 were basalt rather than highland material like that found at Descartes. During their third day on the Moon, Cernan, and Schmitt drove to the base of North Massif where giant boulders had rolled and bounced downhill, leaving distinctive tracks along the way. One such enormous boulder, Split Rock, had broken up into five main pieces. The astronauts spent over an hour, hammering chips and collecting nearby rock and soil samples. Then it was on to the Sculptured Hills, another geological unit at the east end of the valley. Here they found a "pretty" crystalline rock, a piece of plagioclase from the ancient lunar crust. It turned out to be more than 4.3 billion years old.

Time was running out for the last Moonwalkers, and after seven hours, Cernan and Schmitt returned to the LM. Before they climbed inside, Cernan announced to listeners on Earth, "We'd like to uncover a plaque that has been on the leg of our spacecraft... I'll read what that plaque says... 'Here man completed his first exploration of the Moon, December 1972 A.D. May the spirit of peace in which we came be reflected in the lives of all mankind.'"

At 5:55 P.M. EST, on December 14, 1972, *Challenger* blasted off from the Moon. The scene was witnessed by millions of television viewers back on Earth. *Challenger* was reunited with the command ship *America*, and after two more days in lunar orbit, Apollo 17

Apollo 17 splashed down in the Pacific on December 19, 1972. Here the Command Module America *floats in the calm waters of the Pacific after Navy frogmen attached an inflatable collar around its base. The hovering helicopter transferred the astronauts to the* USS Ticonderoga *which is seen in the background.* (NASA)

started its return journey to Earth. During the voyage, Evans retrieved the Scientific Instrument Module (SIM) bay film cassettes.

Like the trip to the Moon, most of the trip home was routine; Apollo 17 was a textbook flight. Splashdown in the Pacific came at 2:25 P.M. on December 19, 1972, ending man's last lunar journey up to this time. But the detailed analysis of the photographs, data, and samples brought home from the Moon had yet to begin.

The Apollo era was over. A centuries-old dream was now a memory.

28 Soviet Moon

While Neil Armstrong and Buzz Aldrin were still on the lunar surface during the historic flight of Apollo 11, a second mystery spacecraft was preparing to land some 700 miles (1100 km) northeast of Tranquillity Base in Mare Crisium (Sea of Crises). The USSR's Luna-15 had been launched atop a Proton rocket on July 13, 1969. It was the first Soviet spacecraft designed to make a genuine soft landing on the Moon. It was also a secret last-ditch attempt to score some sort of propaganda victory by bringing back a small sample of lunar soil before the astronauts did. Luna-15 was a dismal failure. The unmanned ship crashed into the Sea of Crises, its flight completely overshadowed by Apollo 11's epic journey.

The Soviet Union joined the world in congratulating the Americans on their monumental success. At the same time, the Soviets denied that they had lost the Moon Race. They claimed they had never been in it!

When Luna-16 finally did achieve an unmanned sample return mission in September, 1970, Soviet officials argued that sending robots to the Moon was cheaper and safer than risking human lives. They supported this view by pointing to the perilous mission of Apollo 13. Yet, long before the Apollo flights, the Soviet Union had given every indication that a Russian would be the first human being to leave footprints in the lunar soil. Nikita Khrushchev had boasted that a Red Banner would beat the Stars and Stripes to the Moon.

Even after Khrushchev's ouster in 1964, the Soviets had continued to score impressive space achievements, including the first space walk in 1965 and the first robotic landing on the Moon in 1966. At the time, Mstislav Keldysh, president of the USSR Academy of Sciences, called these successes "significant new steps" that opened "great possibilities for future manned flights to the Moon." The cosmonauts themselves also talked about such trips. Yuri Gagarin, the first man in space, predicted that a Soviet research laboratory would be established on the Moon in the 1970s. During a meeting in May, 1967, Pavel Belyayev told astronaut Michael Collins that he (Belyayev) was training for a trip around the Moon "in the not-too-distant future."

Meanwhile, NASA Administrator James Webb repeatedly warned that the Soviets were developing a massive new launch vehicle—bigger than America's giant Saturn V. The purpose of such a booster seemed obvious. With the 50th anniversary of the 1917 Russian Revolution approaching in November, 1967, the Soviet Union might attempt to mark the occasion either by landing a man on the Moon, or sending one around it. However, like the Apollo 1 fire that January, the fatal flight of Soyuz-1 in April, 1967 slowed down the race to the Moon, though neither tragedy stopped it.

Within a year, Apollo was back on track, and so was Soyuz. Indeed, a lunar version of the Soyuz spacecraft was tested several times using the name Zond as a cover. A manned flight around the Moon seemed imminent. It was, but the mission was flown by America's

Luna 16 retrieved the Soviet Union's first lunar sample when it brought three ounces (101 gm) of material back to Earth from the Sea of Fertility in 1970. An automated drill placed the soil into a beach ball-size return capsule like the one seen atop this replica of the Luna 16 spacecraft in a Moscow museum. (Boston Museum of Science)

Apollo 8 in December, 1968. With Apollo 11's landing, seven months later, the Moon Race was over.

In 1989, under Mikhail Gorbachev's new policy of *glasnost* (openness), Radio Moscow broadcast a report, admitting that the Soviet Union had been in the Moon race after all. The report cited the failure to perfect a successful Moon rocket as the main reason the Soviets lost the race.

Boris Belitzky, Radio Moscow's long-time science reporter, and Vladimir Lytkin, Scientific Director of the Tsiolkovsky Cosmonautics Museum in Kaluga, recently provided some insights into the Soviet Moon program.

"There were two parts to [our lunar program]," Belitzky explained. "The first would be a manned circumnavigation of the Moon. This was to be done in a spacecraft of the

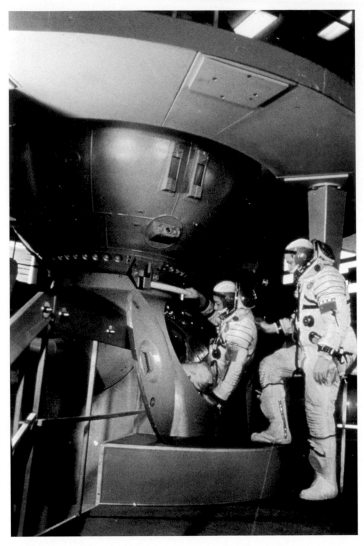

The Soyuz (Russian for "Union") spacecraft was designed for a variety of roles, from shuttling cosmonauts to Earth-orbiting space stations to sending them to the Moon. It is still in service today. Here two cosmonaut trainees enter a Soyuz simulator for a practice space flight. (Novosti Press Agency)

Zond series." Zond was a modified Soyuz designed to send two men around the Moon without going into lunar orbit. The last test mission, Zond-8, was not flown until October, 1970. Coming nearly two years after Apollo 8, it made no sense to continue the effort.

The second part of the Soviet program held more promise. Even if they came in second, the USSR still wanted to put a man on the Moon. Belitzky described "a lunar module with one cosmonaut to make the landing while another cosmonaut remained in a command module in lunar orbit." Like the Americans, Korolev had chosen Lunar Orbit Rendezvous as the best method of reaching the Moon's surface.

According to information provided by Lytkin, the Soviet lander was roughly half the size of the Apollo Lunar Module. Its first unmanned, Earth orbital test flight took place on November 24, 1970, under the name Kosmos-379. Two more tests took place, on February 26, 1971 as Kosmos-398, and on August 12, 1971 as Kosmos-434.

All three lunar lander tests were successful, but the giant N-1 Moon Rocket was problem-plagued from the outset. Lytkin summarized its troubled history. "The first N-1 was launched shortly after noon on February 21, 1969. Ten seconds after lift-off, two of the first stage engines shut down (out of a total of 30). Severe vibrations in the rocket severed a liquid oxygen line, starting a fire. The remaining engines stopped running, and the huge booster fell back to the ground and exploded." The second test, on

Soviet plans called for a Proton booster rocket, similar to this one being erected at the Baikonur Cosmodrome, to carry a pair of cosmonauts around the Moon in a modified Soyuz spacecraft. However, Apollo 8 beat them to it, and the Soviets' plan for a circumlunar flight was quietly abandoned. (Novosti Press Agency)

July 3, 1969, ended the same way. According to Lytkin, "One second after lift-off, an engine blew apart when a small piece of metal became lodged in the oxidizer pump. Once again, the vehicle fell down and exploded, completely destroying itself and the launch pad." The third test came on June 27, 1971. This time, the N-1 made it off the pad, but soon started flying out of control. The vehicle failed after less than one minute of flight. The fourth, and final, N-1 test took place on November 23, 1972, just two weeks before Apollo 17. A modified Soyuz command ship and a lunar lander mock-up sat atop the tall, tapered Moon rocket. The N-1 rose majestically from its Baikonur launch pad, but just 90 seconds after launch, a fuel line ruptured, starting a fire and causing an explosion.

A fifth N-1 could have been ready to fly in the summer of 1974. However, after four failures the N-1 was canceled, along with the entire Soviet manned lunar landing program. The leftover rockets and spaceships were either hidden away, placed into storage, or scrapped. Ironically, 22 years after the N-1's demise, a surplus NK-33 first-stage engine was tested in the United States for possible use in American-built rockets.

Belitzky cited a late start, a shortage of funds, disagreements among the designers, and the lack of a strong guiding hand following Korolev's death as the reasons for the Soviet Union's failure to land a man on the Moon. Belitzky also observed that Korolev's successor, Vasili Mishin, opposed the decision to abandon the Moon program. "He was convinced that if the project had been pursued to its successful culmination, there would already be a research facility operating on the surface of the Moon." This, no doubt, would have generated an American response.

Instead, the Soviets developed their Salyut and Mir Earth orbiting space stations, while the United States pursued its Space Shuttle program. Eventually, the two countries combined their space efforts, starting with the joint Apollo Soyuz Test Project in July, 1975.

Though they lost the race to the Moon, the Soviet Union contributed to man's understanding of the Moon by retrieving their own lunar samples. Three unmanned Luna spacecraft touched down at different sites, each time drilling into the regolith. Luna-16 landed due east of Tranquillity Base in Mare Fecunditatis (Sea of Fertility) on September 20, 1970. Four days later, its return capsule parachuted to a safe landing in Kazakhstan. It carried just 3.5 ounces (101 grams) of mare basalt that had formed some 3.4 billion years ago. Luna-20 touched down on February 21, 1972, in a highland area due north of Luna-16 and near the crater Apollonius. It returned a small sample of anorthositic fragments, as one might expect from a highland landing site. The last lunar lander, Luna-24, reached Mare Crisium on August 18, 1976, and retrieved a 5 foot (3.6 m) core sample that contained mare basalt, roughly 3.3 billion years old. With the delivery to Earth of a few ounces of randomly selected lunar material in their beach ball-sized return capsules, the Luna spacecraft added small, but significant pieces to the lunar jigsaw puzzle.

The Soviets also landed two unmanned rovers on the Moon. Luna-17 carried Lunokhod-1 to the northwestern shore of Mare Imbrium on November 17, 1970. Over the next ten months it traveled 6.5 miles (10.5 km) across the Moonscape, sending back television images and soil measurements. Luna-21 took Lunokhod-2 to the crater Le Monnier, 110 miles (180 km) due north of Taurus-Littrow, on January 16, 1973. The rover traveled 23 miles (37 km) over four months. Two more Lunas took photos from lunar orbit in 1971 and 1974.

Although their scientific return was rather limited, compared to Apollo, these unmanned Luna probes were harbingers of future robotic expeditions to the Moon and Mars.

The Soviet Union sent two unmanned Lunokhod rovers to the Moon in 1970 and 1973. Here is a replica of Lunokhod-2. Eight wheels driven by electric motors carried the vehicle 23 miles (37 km) over a four-month period. Power was provided by solar cells. Cameras helped Earthbound controllers steer a path around rocks and craters. (Boston Museum of Science photograph)

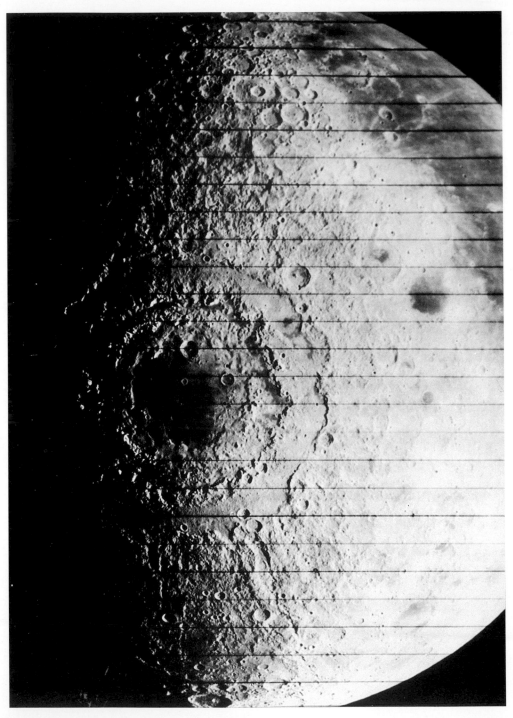

The asteroid impact that created the multi-ringed, 600 mile (965 km) diameter Orientale Basin ended the Early Imbrian Epoch. This photograph was taken by Lunar Orbiter 4 as it circled just beyond the Moon's left edge as seen from Earth. Long grooves cut by flying debris extend radially from the center of the basin. The dark patch on the upper right edge of the Moon on this Lunar Orbiter 4 photograph is Oceanus Procellarum (Ocean of Storms). (NASA)

Part 3

The New Moon

29   Where Did the Moon Rocks Come From?

How did scientists determine how the Moon rocks were formed and the age of each sample? As we have seen, space debris — meteoroids, asteroids, and comets — played a major role in shaping the Moon's surface. To better understand this process, let's consider some down-to-Earth analogies.

Imagine throwing rocks onto a freshly-hardened concrete highway, each rock making a small dent, or crater, in the pavement. If we keep throwing rocks, day after day, week after week, we would create our own little Moonscape, across which few cars could comfortably travel. If we imagine our highway as the Moon's original hardened crust, and our rocks as meteorites in assorted shapes and sizes, then we have a good picture of how the lunar surface was formed.

Each time the Moon was struck by space debris, the crustal material was broken up, a crater was formed, and ejecta was splattered radially around the crater's rim. Moreover, this ejecta created secondary impacts, forming additional small craters when the ejected material fell back to the surface. The same thing happens when you slam a large rock into soft mud. *Splash!*

Now imagine this process taking place over and over again throughout 4½ billion years. The term "gardening" is used to describe the continual shattering and overturning of lunar crust to form the regolith, a top layer which consists of broken material ranging in size from giant boulders to grains of sand. While there is no wind or water on the Moon to erode its surface, the steady bombardment over the eons of meteorites, large and small, has had the same effect, pulverizing the crust and wearing down the edges of rocks and craters. This erosion of lunar craters enabled astrogeologists to describe the Moon's geologic history a decade before the first Apollo landing missions. The scientists studied the Moon's surface by peering through telescopes and by carefully examining detailed, lunar photographs. Then they applied the analytical techniques used in a branch of geology, called stratigraphy.

Here on Earth, we can see how different layers of rock — strata — record different eras in our planet's history. The Grand Canyon, for example, reveals layer upon layer of rocks, spanning hundreds of millions of years of Earth's past. Younger layers sit atop older layers. Primeval plant and animal fossils can be found in rock layers buried far below the rock layers that record more recent life forms. (Except for the 12 Apollo astronauts, there has never been any life on the Moon, so there were no fossils among the Moon rocks.)

Dr. Eugene Shoemaker used stratigraphy to produce the first geologic maps of the Moon, showing the surface distribution of various rock types, age relationships, and structural features. Here's how lunar stratigraphy works: Start with an airplane, a supply of bombs of various sizes, and a large freshly-plowed field. In the spring of Year One, you fly over the field, drop a few bombs, then view the results from the air. You'll see several, sharply-defined craters. The larger ones, produced by the bigger bombs, have raised rims and rayed patterns of ejecta, debris thrown out by each blast, surrounding them.

In Year Two, after wind, snow, and rain have had a chance to erode the craters, they appear less well-defined. The distinctive ray patterns are no longer as obvious. Onto this

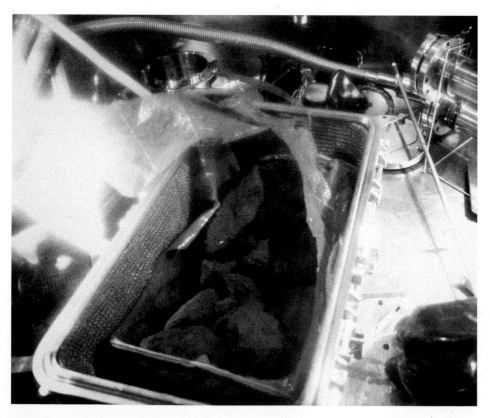

The first Moon rocks brought back to Earth were these samples collected by Apollo 11 astronauts Neil Armstrong and Buzz Aldrin. Here the rock box in which they were transported is carefully opened in the controlled environment of the Lunar Receiving Laboratory at the Manned Spacecraft Center in Houston. (NASA)

Apollo 16 sample Number 68815 was a breccia containing numerous pieces of white feldspar-rich rock. The dark cavities are vesicles, formed as bubbles of gas were released when a meteorite strike produced the molten material that formed this rock. (NASA)

altered surface, you drop some more bombs. Once again, sharply-defined craters are formed, some of them overlapping the older craters that were produced in Year One.

In Year Three, the craters generated in Year Two are now eroding, while those produced in Year One are even less distinct. You drop another series of bombs. Now, as you observe the surface below your airplane, it has become easy to distinguish the relative ages of the various craters. Those that are sharply-defined are the most recent, while those which are overlapped and eroded, are older. The craters which are the least apparent are from Year One, and are clearly the oldest. Using a similar technique, Shoemaker and his colleagues were able to analyze the Moon's features and determine the different stages in the Moon's battered history (see Appendix C).

One of the most obvious features on the lunar surface, as seen from Earth, is Mare Imbrium (the Sea of Rains). It was used as a benchmark in analyzing the Moon's history. You might compare it to Year Two in our bomb field analogy. Four, easily-recognizable, stratigraphic divisions were identified on the Moon. Features created by the giant impact that formed the Imbrium basin were described as "Imbrian" on lunar geologic maps. Older formations were termed "Pre-Imbrian," while more recent landmarks were

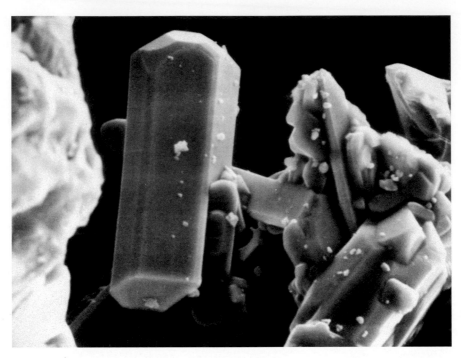

Under an electron microscope's scrutiny, crystals of the mineral apatite—calcium phosphate—are visible in a small cavity of a breccia collected by the Apollo 14 astronauts. Formed shortly after the giant impact that created the Imbrium Basin 3.85 billion years ago, the crystals were propagated from hot vapor as the rock cooled. (NASA)

categorized by the slightly-eroded crater, Eratosthenes, and the fresh-looking crater, Copernicus.

The Apollo landing sites were chosen so that representative samples from each of the different features could be collected, thus enabling scientists to assign dates to the times of their formation. Apollo 14 samples were used to date the Imbrium Basin at 3.85 billion years, while Apollo 12 samples dated Copernicus at a mere 800 million years. Other samples identified the ages of other features. The position and size of lunar rocks and craters gave clues to their origins, whether from a small nearby crater, or from a huge depression halfway around the Moon. The cluster of rocks and secondary craters on the valley floor at Taurus–Littrow, for example, was traced all the way back to the large crater, Tycho.

Dating the Moon rocks was relatively easy. By measuring the decay of radioactive elements in each sample, its age could be determined in the same manner Earth rocks are dated. We now know that some of the youngest features on the Moon pre-date the dinosaurs by hundreds of millions of years, while some of the oldest Moon rocks are nearly as ancient as the Solar System itself.

30 What We Found There

Between 1959 and 1976, the United States and Soviet Union launched a virtual flotilla of spaceships to explore the Moon. The six Apollo lunar landing missions returned 844.5 pounds (383.8 kg) of lunar samples to Earth. Three unmanned Luna probes brought back another 10.5 ounces (300 grams) of Moon material. Although the Soviet robots retrieved only a handful of lunar samples compared to the large collection gathered by the American astronauts, they nevertheless made significant contributions to lunar science, as their samples came from regions of the Moon the astronauts had been unable to explore. What did the samples tell us?

To begin with, we discovered that the Moon is similar, in some basic ways, to Earth. The minerals found in lunar rocks are the same ones found in terrestrial rocks, although in different combinations and proportions. We also learned that Moon rocks are igneous,

The mountains along the horizon in this Apollo 8 photograph are part of the ancient, pre-Nectarian South Pole-Aitkin basin. This huge far side basin, like its near side counterpart, the Procellarum basin, has been virtually erased from view by impacts during the Moon's early history. (NASA)

The ancient crater, in the foreground of this photo from Apollo 8, is Goclenius, on the western edge of Mare Fecunditatis (Sea of Fertility). This lava-filled crater's floor is crisscrossed by cracks and fissures which are likely related to the uplifting of the crater floor, perhaps by lava from below; and by linear rilles, which were formed when linear sections of crust dropped down along parallel faults. Since they are superimposed on other features, these rilles must be relatively young. (NASA)

that is, they were formed when hot magma—liquid rock—cooled, crystallized, and solidified, just like igneous rocks on our own planet. Like Earth, the Moon is not a homogeneous body, but is composed of a variety of different rock types.

As we have already seen, most of the rocks collected at the mare sites were basalts—dark fine-grained rocks, rich in iron and titanium. They were formed by successive lava flows that filled the Moon's giant basins between 3.8 and 3.2 billion years ago. The highland rocks were largely anorthosites—older and lighter than mare rocks, rich in aluminum and feldspar. They were formed as part of the original lunar crust between 4.5 and 4.0 billion years ago. A third group of rocks, falling between the basalts and anorthosites in age, are rich in potassium (K), and contain unusually high amounts of rare-earth elements (REE), and phosphorus (P). They are basaltic rocks, dubbed "KREEP," and were found at the Apollo 12 and Apollo 14 landing sites. KREEP is the most highly radioactive material found on the Moon. Its exact origin is still something of an enigma.

Many of the rocks brought to Earth by the Moon missions were breccias. As briefly described in previous chapters, breccias were formed when two or more pre-existing rocks were fused together by a meteorite strike. The enormous heat energy released by the impact melted some of the lunar material and splattered a mixture of rocks and molten matter across the Moon's surface. As this mixture cooled, rocks and angular rock fragments were bonded together, creating breccias. Some of the molten matter also solidified as tiny beads of glass, resembling microscopic marbles, which explains why natural glass is quite common on the Moon. (You may recall that the orange soil found at Shorty Crater by the Apollo 17 astronauts originated as small beads of orange glass which had formed in a volcanic fire fountain. Their color was due, in part, to their sulfur content.)

Another key discovery was that space debris — not volcanism — was primarily responsible for the countless lunar craters. The Moon has endured a violent past.

Prior to the Space Race, we knew very little about the Moon and planets. A wall separated astronomy from the earth sciences. The nature of alien worlds, including the Moon, was largely unknown, and many scientists considered those worlds unrelated to our own planet. The Moon belonged to the science of astronomy, while Earth belonged to geology. The two scientific disciplines had failed to come together in a meaningful way, and as a result, our connection to the rest of the Solar System wasn't always generally understood or fully appreciated.

The Moon missions radically changed this perspective. For the first time in human history, Earth could be compared to another body in space, giving birth to a new science — comparative planetology. The field was founded by a small group of insightful theorists — one of whom was Dr. Eugene Shoemaker — who recognized the influence that extraterrestrial events have had on our own environment. These scientists believed that one key to fully understanding Earth was to learn the history of its neighbors. Thus, they regarded Moon rocks as potential cosmic Rosetta stones, which could help us decipher the Solar System's mysterious ancient past.

Now, four decades after the dawn of the Space Age, the lunar surface bears human footprints, Martian soil has been scooped up and analyzed by robotic Viking landers, the exotic surface of Venus has been thoroughly mapped by Magellan, the Moon-like face of Mercury has been revealed by Mariner 10, and the giant outer planets and the moons that orbit them have been explored by the Voyager and Galileo probes. We have also seen

Saturn's magnificent rings, active volcanoes on Io, and icebergs on Europa (moons of Jupiter), and the peanut-shaped nucleus of Halley's Comet.

Our new-found knowledge of the Moon has helped us to interpret some of the features we have seen on these other worlds, and given us a cosmic yardstick to measure the approximate age and composition of rocky planets, such as Mercury and Mars. Knowing how the size and appearance of lunar craters are related to their age and composition, we can apply some of the same criteria to the craters we have observed on other worlds, including Earth.

The Moon's history suggests that a steady rain of space debris fell on all the planets during the Solar System's early years between 4.6 and 4.0 billion years ago. The most intense period of bombardment ended some 3.8 billion years ago. Few large objects have struck the Moon (or Earth) during the past 1.0 billion years, but major impacts still occur on occasion. An asteroid hitting Earth could have been responsible for the extinction of the dinosaurs 65 million years ago. In 1908, a small comet fragment or asteroid exploded in the atmosphere over Siberia. The shock wave produced by the explosion was 2,000 times more powerful than the atomic bomb which destroyed the Japanese city of Hiroshima at the end of World War II. The Tunguska event, as the Siberian blast is called, leveled forests and incinerated wildlife for miles around. Had the event occurred over a major city, hundreds of thousands of people would have been killed.

Fortunately, the absence of any large fresh craters on the lunar surface tells us that the era of giant impacts is long past. But while the chances of a life-threatening cosmic collision with Earth appear to be remote, the fiery crash of Comet Shoemaker-Levy 9, (co-discovered by Eugene Shoemaker, his wife Carolyn, and David Levy) into Jupiter in 1994 was a vivid reminder that such a cataclysm is still possible. Had Comet Shoemaker-Levy 9 struck Earth, its impact would have produced a global catastrophe. For that reason, Shoemaker, one of the pioneers of lunar research, has helped to pioneer a new NASA program which searches the skies for cosmic threats. We have no way of knowing whether a collision will occur next week, next month, or a million years from now. If an Earthbound comet or asteroid is detected early enough, it could conceivably be destroyed or deflected with an armada of long-range missiles and nuclear warheads.

The Moon missions taught us a great deal about our Solar System's violent past. That knowledge, together with the leftover weapons of the Cold War, might some day be used to safeguard Earth's future.

31 The Cost and the Legacy

The cost of lunar exploration was a staggering $25 billion—over $100 billion in 1997 dollars—for the U.S. Moon missions, plus untold billions of Russian roubles. Placing an exact price on the Soviets' efforts remains problematic, as many aspects of their lunar program remain a secret. But it is certainly safe to say that the Soviet Union invested enormous amounts of its own resources in the race to the Moon.

The Moon exacted a human price also. Gus Grissom, Ed White, and Roger Chaffee lost their lives in the Apollo 1 fire, while Vladimir Komarov lost his life aboard Soyuz-1. The frantic pace of the Moon Race also took its toll on the men and women who worked as scientists, engineers, and technicians. No one will ever know how many stomach ulcers, heart attacks, and broken homes occurred as a result of the Moon missions, but those who survived say the number was high.

Was going to the Moon worth the cost?

The answer is necessarily somewhat subjective. At the time the decision was made to go to the Moon, national prestige and political ideology were clearly on the line. Gov-

America wins the Moon Race and senior NASA officials celebrate the successful completion of the first lunar landing in July, 1969. This scene in the Mission Operations Control Room at the Manned Spacecraft Center in Houston was repeated in countless other homes and offices all across the nation. (NASA)

ernment leaders recognized the propaganda value of being the first nation to land a man on the Moon. The first lunar landing was a dazzling technological achievement, but the Moon's luster rapidly faded. People were concerned with their own, earthly problems, and politicians concluded that lunar rocks had no lasting effect on the lives of ordinary citizens. Moon missions would not end wars, cure cancer, or conquer poverty.

An estimated 30,000 new products originated in the space program, including: new reading machines for the blind, new navigation systems for jetliners, new fireproof suits for firefighters, as well as many other items for industrial and consumer use. Advances in communications and weather forecasting can be enjoyed each night on the evening news. Today's computer and electronic industries, which provide millions of jobs around the world, owe their dynamic growth in large measure to the American investment in Project Apollo. Schools and universities, as well as the medical profession, have reaped enormous benefits from Space Age technology.

In short, the space program has stimulated the American and global economies through the development of advanced technologies, which in turn have served the average citizen in ways too numerous to count. From a purely economic standpoint, one can argue that Apollo was certainly worth its $25 billion price tag.

Still, some people place a value on things which cannot be measured in dollars and kopecks, or immediate tangible terms. They are the visionaries who cherish exploration and basic scientific research, knowing that the New Frontiers of today, might become the New Worlds of tomorrow. John Kennedy, a dynamic, young president, demonstrated such vision when he challenged the American people to boldly set sail on the "new ocean" of space.

What about the opinion of the American public? In a 1970 poll, 39 percent of the American people said that going to the Moon had been worth the cost, while 56 percent said it had not. In 1991, 56 percent of the American people felt that the money invested in space exploration could be better spent on education and health care. Thus, over 21 years the percentage of taxpayers opposed to spending money on space exploration remained unchanged. Still, space proponents have always argued that the space program's costs have been greatly outweighed by its economic benefits.

Some of the most memorable images taken by the Apollo astronauts were of Earth, and they gave us an appreciation of the fragile nature of our own planet. The Apollo Program gave new meaning to the environmental movement and made people more aware of the need to preserve our precious resources.

From a scientific perspective, the Apollo Program gave us a new understanding of the Solar System and our place in it. Over the past few decades, the mysteries and wonders of the cosmos have unfolded at an astounding rate. Apollo opened a door to the Universe—and to our future. We humans are a curious species, having always wondered about the nature of things. Apollo was humankind's first visit to another world, and as such, it was a beginning, not an end. Apollo served as a demonstration of the technology required for even bolder journeys to Mars and beyond. The Moon missions were, in effect, our first tentative steps toward the stars. As with most pioneering efforts, the true value of those steps will only become clear with the passage of time.

In the wake of their defeat, the Soviets swallowed their pride and pursued a new course of action. Cooperation slowly replaced competition. The last Apollo flight was, appropriately, the first joint mission in space. On July 17, 1975, three Apollo astronauts linked up with two Soyuz cosmonauts in Earth orbit. The Apollo Soyuz Test Project was criticized by some as an expensive political stunt, but the historic flight laid the foundation for a more genuine space partnership with Russia. A series of joint missions with the American Space Shuttle and Russian Mir space station launched in 1986, have paved the way for a more challenging undertaking—the construction of an International Space Station in Earth orbit by the turn of the century. The former space rivals are also working together on a series of unmanned missions to explore Mars.

The exploration of the Moon gave birth to a new science—comparative planetology. This battered, Moon-like planetary surface is Mercury, as photographed by Mariner 10 during its swing by the closest planet to the Sun. Studying lunar history and geology has helped scientists to better interpret some of the features observed on other planets throughout the Solar System—including our own. (NASA)

32 Looking to the Future

When the first people trekked across the Bering Strait landbridge from Asia to the Americas, 28,000 years ago, they were probably searching for richer hunting grounds. When the Vikings explored North America, 1,000 years ago, they were likely seeking timber for shipbuilding. When Columbus "discovered" the New World, 500 years later, he was expecting to find a shorter sea route between Europe and Asia. Economics, not science, was the stimulus that drove each of these great human adventures, and economics will likely determine when humans go back to the Moon.

The Cold War, which was the impetus behind the first Moon missions, now belongs to history, and so do the old political and military motives for going to the Moon. An economic justification for the trip will be necessary for humans to return to the Moon. After all, if Columbus had not brought gold back from the New World, would other ships have gone?

The Apollo astronauts did not discover gold nuggets on the Moon. However, they did find a potentially priceless economic resource in the ancient lunar soil. In the view of scientist-astronaut, Dr. Harrison Schmitt, "Maybe the most important thing that happened as a result of the Apollo Program wasn't realized until late 1985. That is when scientists at the University of Wisconsin in Madison recognized that we had found a significant concentration of a light isotope of helium, called helium-3, in the lunar soils. And the reason this discovery was significant is because helium-3 constitutes probably the most ideal fuel for fusion power that we could find."

Fusion reactors, when perfected, will revolutionize the energy industry. In a fusion reaction, atomic nuclei are combined to form more massive nuclei, releasing energy in the process. Fusion is what powers the stars, including our own Sun. The tremendous heat and pressure at the Sun's core is great enough to fuse hydrogen atoms into helium atoms through a complex series of reactions that release enormous amounts of energy. Fusion gives hydrogen bombs their awesome power. The present challenge is to harness that power and convert it into electricity. It's an engineering problem that will likely be solved within the next decade.

Once fusion reactors are developed, Schmitt and others believe that, because of its benign nature, helium-3 will make an ideal fuel. Unlike other potential fuels, helium-3 will not produce the radioactive waste which is associated with today's nuclear power plants. Helium-3 is extremely rare on Earth, but it is plentiful on the Moon. In fact, some scientists have estimated that enough helium-3 exists in the upper few yards (meters) of the lunar regolith to supply our planet's energy needs for hundreds, perhaps thousands of years.

"Someday," Schmitt asserts, "it is almost certainly going to be economically and environmentally sound to extract helium-3 from the lunar soils, bring it back to Earth, and use it to provide a viable alternative to fossil fuels for electrical power generation."

HERE MAN COMPLETED HIS FIRST
EXPLORATIONS OF THE MOON
DECEMBER 1972, A.D.
MAY THE SPIRIT OF PEACE IN WHICH WE CAME
BE REFLECTED IN THE LIVES OF ALL MANKIND

EUGENE A. CERNAN
ASTRONAUT

RONALD E. EVANS
ASTRONAUT

HARRISON H. SCHMITT
ASTRONAUT

RICHARD NIXON
PRESIDENT, UNITED STATES OF AMERICA

This plaque was left on the Moon by the last Apollo astronauts. Mounted to the forward leg of the Lunar Module descent stage, it marked the end of a beginning. (NASA)

Mining helium-3 on the Moon would require the development of a lunar base and a helium-3 processing plant. The actual mining operation would begin with lunar bulldozers scooping up bulk quantities of regolith. This material would then be hauled to a nearby processing plant where the helium-3 would be extracted from the soil by heating it to approximately 1470 degrees Fahrenheit (800 degrees Centigrade). Next, the processed fuel would be loaded into a cargo ship, and transported from the lunar surface to an Earth-orbiting space station. Finally, it would be transferred to a Space Shuttle, brought to Earth, and delivered to fusion power plants around the globe.

The process of extracting the helium-3 from the lunar soil, Schmitt adds, "would also yield several by-products, such as hydrogen and water, which is exactly what you need in order to begin the settlement of our Solar System." Hydrogen could be used to produce rocket fuel, and water would help to sustain life on the Moon. Oxygen could also be extracted from the lunar soil.

Schmitt believes a profitable lunar mining operation could be established by the year 2015. The cost of building the lunar infrastructure, he maintains, would be less than that of building the Alaskan oil pipeline. The private sector would take the lead in this venture. Schmitt has proposed an initiative dubbed, INTERLUNE (INTERnational LUNar Enterprise), which would call for the creation and sale of helium-3-based fusion technology. INTERLUNE would spearhead the development of the lunar resource production, sales, and distribution system.

The Space Shuttle Columbia *touches down in California. The NASA program may be a harbinger of a future era, when shuttles carry cargo — for example, helium-3 — back to Earth, as well as satellites and supplies into space from Earth's surface.* (NASA)

Dr. Sally K. Ride, another scientist-astronaut and America's first woman in space, has offered another plan for resuming lunar exploration. Dr. Ride produced a report for NASA in 1987 entitled "Leadership and America's Future in Space." The so-called Ride Report charted a course for NASA's future, and called for the adoption of four key initiatives:

- An integrated approach to observing our own planet, using Earth-orbiting satellites.
- The continued exploration of other bodies in the Solar System (planets, asteroids and comets), using unmanned spacecraft.
- A manned expedition to Mars.
- An outpost on the Moon.

"Although explorers have reached the Moon," Dr. Ride noted, "the Moon has not been fully explored." Her "Outpost on The Moon" initiative suggests a three-phase approach:

Phase I. Robotic spacecraft map the Moon's surface in great detail, and identify potential sites for a lunar base.

Phase II. Small crews land on the Moon, each crew staying for one to two weeks. The astronauts would set up scientific instruments, construct research and habitation modules, and build a pilot plant for extracting lunar oxygen.

Phase III. A permanent outpost is established on the Moon, with up to 30 people "productively living and working on the lunar surface for months at a time."

Developing the technology for living on another world would facilitate the eventual establishment of an outpost on Mars. Ride also identified several scientific reasons for going back to the Moon, one of which is to further explore the lunar surface and to prospect for resources while learning more about the Moon's geologic history. The Moon would also make an excellent platform from which to study the Universe. Optical telescopes would be free of the atmospheric effects which plague terrestrial astronomers, and radio telescopes on the Moon's far side would be shielded from the electromagnetic noise generated on Earth. Ride's proposed lunar outpost might serve as a catalyst for the sort of industrial enterprise advocated by Schmitt. "This initiative," Ride wrote, "would push back frontiers, not to achieve a blaze of glory, but to explore, to understand, to learn, and to develop."

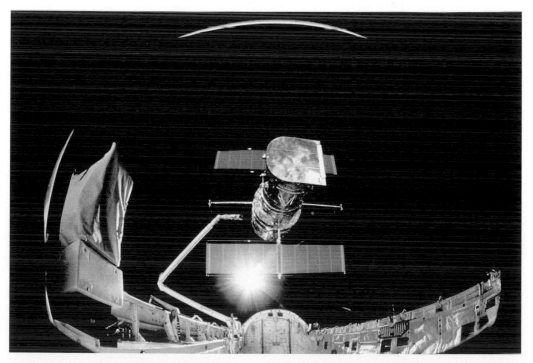

While the Hubble Space Telescope (HST) has provided many new insights to astronomers, its operation is restricted by its 90-minute orbit. Only half that time is spent in darkness, and since the telescope is moving, it must acquire, and track, its target on each orbit. The Moon would provide an extremely stable telescope platform, and since darkness lasts a full two weeks, there would be large blocks of time in which to make observations. Here the HST is deployed from the Space Shuttle Discovery *on April 25, 1990.* (NASA)

While most politicians are unwilling to commit scarce tax dollars to such bold initiatives, lunar exploration is being resumed on a more modest scale. Both Japan and the United State have plans to send probes to the Moon in 1997. The U.S. probe, NASA's Lunar Prospector, will orbit the Moon from pole to pole, giving scientists new data about lunar resources, including possible ice deposits in permanently shadowed regions at the poles. (In 1994, another American spacecraft, Clementine 1, returned some two million detailed images of the Moon's surface from lunar orbit.) We still have much to learn about our nearest neighbor, including the exact nature of its core.

Since 1996, when NASA announced the tentative discovery of fossil life on Mars, the world's attention has been focused on the Red Planet. Over the next decade, a new

Sergei Korolev (1906-1966) was the Soviet Union's famous and long anonymous "Chief Designer of Rockets and Space Systems." Korolev was largely responsible for the succes of his nation's early space projects, including Sputnik, Luna, Vostok, and Soyuz. (Tsiolkovsky Museum via V. Lytin.)

Shortly before the flight of Apollo 11, Wernher von Braun posed in front of the giant Saturn V Moon Rocket. As a teenager, von Braun dreamed of sending rockets to the Moon. (NASA)

flotilla of unmanned spacecraft will be sent to the rust-colored Martian surface in an attempt to answer that most tantalizing of all questions: *Are we alone?* Eventually, humans will make the journey to Mars, and the Moon, with its reduced gravity, could become the main port for voyages to that planet. Dr. Schmitt calls the Moon "a natural space station." He looks forward to the day when men and women return to the Moon to live, to work, and to explore. When will that day come? No one can say for certain as there are no current plans to send humans to the Moon.

It's been 25 years since Jack Schmitt and Gene Cernan left their footprints in the lunar soil. In a larger sense, those footprints were put there by the imagination of men like Jules Verne, Konstantin Tsiolkovsky, Robert Goddard, Sergei Korolev, Wernher von Braun, Eugene Shoemaker, Nikita Khrushchev, and John Kennedy. Every great adventure is launched by imagination.

As Albert Einstein so wisely observed, "Imagination is more important than knowledge."

This map of the Moon's near side shows the six Apollo landing sites. (NASA)

Appendix A: *The Moon's Vital Statistics*

Diameter .... 2,160 miles (3,474 km)

Circumference at the equator .... 6,785 miles (10,920 km)

Minimum distance to the Moon .... 228,000 miles (364,800 km)

Maximum distance to the Moon .... 252,000 miles (403,200 km)

Average distance to the Moon .... 238,865 miles (384,400 km)

Orbital period around Earth .... 27 days, 7 hours, 43 minutes

Synodic Period .... 29 days, 12 hours, 44 minutes
(Full Moon to Full Moon)

Surface area covered by maria .... 16.9 percent of total surface area
31.2 percent of Near Side
2.6 percent of Far Side

Total area visible from Earth: .... 59.0% (owing to the Moon's librations)

Maximum daytime temperature .... +243 degrees F (+130 degrees C)

Minimum nighttime temperature .... –261 degrees F (–162 degrees C)

Moon's surface gravity .... 0.17 or one-sixth of Earth's gravity

The Apollo 8 crew stands in front of a Command Module simulator, prior to their Christmas voyage to the Moon. Standing (left to right) are: Command Module Pilot James A. Lovell, Jr.; Lunar Module Pilot William A. Anders; and Commander Frank Borman. (NASA)

Appendix B: *The Apollo Lunar Missions*

Apollo 8

Commander: Frank Borman
Command Module Pilot: James A. Lovell, Jr.
Lunar Module Pilot: William A. Anders

Launch Vehicle: SA-503
Command Module: CSM-103
Lunar Module: none

Launch date: December 21, 1968
Lunar touchdown: none
Splashdown on return to Earth. December 27, 1968
Flight time: 147 hours, 0 minutes, 42 seconds

Notes: Apollo 8 was the first manned spacecraft to leave Earth's gravity and orbit the Moon (10 revolutions). The crew took detailed photographs of the lunar surface for future landing missions.

Apollo 10

Commander: Thomas P. Stafford
Command Module Pilot: John W. Young
Lunar Module Pilot: Eugene A. Cernan

Launch Vehicle: SA-505
Command Module: CSM-106 *Charlie Brown*
Lunar Module: LM-4 *Snoopy*

Launch date: May 18, 1969
Lunar touchdown: none
Splashdown on return to Earth: May 26, 1969
Flight time: 192 hours, 3 minutes, 23 seconds

Notes: Apollo 10 tested the Lunar Module in lunar orbit (31 revolutions) without actually landing on the Moon. It was a "dress rehearsal" for Apollo 11.

Apollo 11

Commander: Neil A. Armstrong
Command Module Pilot: Michael Collins
Lunar Module Pilot: Edwin E. Aldrin, Jr.

Launch Vehicle: SA-506
Command Module: CSM-107 *Columbia*
Lunar Module: LM-5 *Eagle*

Launch date: July 16, 1969
Lunar touchdown: July 20, 1969 Sea of Tranquillity
Splashdown on return to Earth: July 24, 1969
Flight time: 195 hours, 18 minutes, 35 seconds

Notes: Apollo 11 was the first manned lunar landing. Stay time on the surface was 21 hours, 36 minutes, 21 seconds. Armstrong and Aldrin collected 48.5 lb (22 kg) of lunar samples during their single EVA.

Apollo 12

Commander: Charles Conrad, Jr.
Command Module Pilot: Richard F. Gordon, Jr.
Lunar Module Pilot: Alan L. Bean

Launch Vehicle: SA-507
Command Module: CSM-108 *Yankee Clipper*
Lunar Module: LM-6 *Intrepid*

Launch date: November 14, 1969
Lunar touchdown: November 19, 1969 Ocean of Storms
Splashdown on return to Earth: November 24, 1969
Flight time: 244 hours, 36 minutes, 25 seconds

Notes: Conrad and Bean spent 31 hours, 31 minutes on the lunar surface. During two EVAs they collected 75.0 lb (34 kg) of lunar samples, plus the television camera and several other pieces from the unmanned Surveyor 3 spacecraft. They also deployed the first Apollo Lunar Surface Experiments Package (ALSEP) on the Moon.

Apollo 13

Commander: James A. Lovell, Jr.
Command Module Pilot: John L. Swigert, Jr.
Lunar Module Pilot: Fred W. Haise, Jr.

Launch Vehicle: SA–508
Command Module: CSM–109 *Odyssey*
Lunar Module: LM–7 *Aquarius*

Launch date: April 11, 1970
Lunar touchdown: none
Splashdown on return to Earth: April 17, 1970
Flight time: 142 hours, 54 minutes, 41 seconds

Notes: An explosion in a Service Module oxygen tank on the way to the Moon crippled the command ship, aborted the mission, and forced the astronauts to use the lunar module as a "life boat" for their return to Earth.

Apollo 14

Commander: Alan B. Shepard, Jr.
Command Module Pilot: Stuart A. Roosa
Lunar Module Pilot: Edgar D. Mitchell

Launch Vehicle: SA–509
Command Module: CSM–110 *Kitty Hawk*
Lunar Module: LM–8 *Antares*

Launch date: January 31, 1971
Lunar touchdown: February 5, 1971 Fra Mauro
Splashdown on return to Earth: February 9, 1971
Flight time: 216 hours, 1 minute, 57 seconds

Notes: Shepard and Mitchell spent 33 hours, 31 minutes on the lunar surface. During two EVAs they deployed an Apollo Lunar Surface Experiments Package (ALSEP), trekked to very near the rim of Cone Crater, and collected 96.0 lb (43 kg) of lunar samples.

Apollo 15

Commander: David R. Scott
Command Module Pilot: Alfred M. Worden
Lunar Module Pilot: James B. Irwin

Launch Vehicle: SA-510
Command Module: CSM-112 *Endeavour*
Lunar Module: LM-10 *Falcon*

Launch date: July 26, 1971
Lunar touchdown: July 30, 1971 Hadley-Apennine
Splashdown on return to Earth: August 7, 1971
Flight time: 295 hours, 11 minutes, 53 seconds

Notes: The first of the extended "J" missions, Scott and Irwin spent 66 hours, 55 minutes on the Moon. During three EVAs they deployed an Apollo Lunar Surface Experiments Package (ALSEP), collected 170 lb (77 kg) of samples, and drove the first car (Lunar Roving Vehicle) on the Moon.

Apollo 16

Commander: John W. Young
Command Module Pilot: Thomas K. Mattingly II
Lunar Module Pilot: Charles M. Duke, Jr.

Launch Vehicle: SA-511
Command Module: CSM-113 *Casper*
Lunar Module: LM-11 *Orion*

Launch date: April 16, 1972
Lunar touchdown: April 20, 1972 Descartes Highlands
Splashdown on return to Earth: April 27, 1972
Flight time: 265 hours, 51 minutes, 5 seconds

Notes: Young and Duke spent 71 hours, 2 minutes on the Moon. During three EVAs they deployed an Apollo Lunar Surface Experiments Package (ALSEP), drove another "Rover," and collected 213 lb (96 kg) of lunar samples.

Apollo 17

Commander: Eugene A. Cernan
Command Module Pilot: Ronald E. Evans
Lunar Module Pilot: Harrison H. Schmitt

Launch Vehicle: SA-512
Command Module: CSM-114 *America*
Lunar Module: LM-12 *Challenger*

Launch date: December 7, 1972
Lunar touchdown: December 11, 1972 Valley of Taurus-Littrow
Splashdown on return to Earth: December 19, 1972
Flight time: 301 hours, 51 minutes, 59 seconds

Notes: longest and last Apollo lunar mission, Cernan and Schmitt spent 75 hours on the Moon's surface. During three EVAs they collected 243 lb (110 kg) of lunar samples, deployed the fifth and final Apollo Lunar Surface Experiments Package (ALSEP), and drove their Rover a total of 22 miles (35 km) around a spectacular valley site.

Also:

After Apollo 17, three more Apollo spacecraft were launched atop Saturn 1B launch vehicles into Earth orbit as part of the Skylab space station program. The last Apollo flight was launched on July 15, 1975. It rendezvoused and docked in Earth orbit with a Soviet Soyuz spacecraft during the Apollo Soyuz Test Project. Apollo 7 in October 1968, and Apollo 9 in March 1969, were flown in conjunction with the lunar program, but tested the spacecraft hardware in Earth orbit.

A total of 24 Apollo astronauts flew to the Moon. Twelve of them walked on its surface.

A nearly Full Moon, unlike any seen from Earth, was captured in this trans-Earth photograph taken by Apollo 16's SIM bay mapping camera. The right half of the picture shows the far side of the Moon. The left half covers the right edge of the near side as seen from Earth. The dark elliptical spot on the left limb is Mare Crisium. The battered face of the far side is nearly devoid of maria, and its heavily cratered surface is representative of the greater portion of the Moon's total surface area. (NASA)

Appendix C: *Lunar Geologic History*

The Moon missions gave us the pieces to a cosmic jigsaw puzzle. Some of the pieces were missing, but we have enough of them to put together a general picture of lunar geologic history. Based on their studies of the Apollo and Luna rock and soil samples, scientists were finally able to match dates with events, dividing the Moon's history into the following time units:

Pre-Nectarian Period   The exact origin of the Moon remains unclear. Most scientists believe it was formed in the early days of the Solar System—about 4.6 billion years ago. In that distant time, particles of rock and iron began to form out of the giant cloud of dust and gas that circled our newborn Sun. These particles flew into each other, creating larger particles which grew bigger and bigger until they formed rocks. Then the rocks slammed into each other, forming boulder-sized bodies. Eventually, the gravitational attraction of the bigger bodies swept up the smaller ones through a process called accretion. As they grew ever larger, these bodies formed protoplanets. All of the rocky inner planets developed in this fashion. The heat generated by the impacting bodies was great enough to melt the young protoplanets—including the Moon—turning them into spherical blobs of molten rock and metal. While the interior of the Moon slowly cooled, its outer layers were covered by a deep magma "ocean."

After 100 million years had passed, the hot magma slowly started to cool, forming the lunar crust. Various minerals began crystallizing, the lighter ones forming anorthositic rocks that rose to the surface, while the heavier ones, such as olivine, sank toward the interior. Curiously, the crust is not uniform. It is somewhat thicker on the far side. There was also a great deal of churning and mixing during this period as a result of convection, the same process that occurs in a pot of boiling soup. Huge meteorites continued to bombard the surface, although the impact rate had decreased significantly from the earliest days. Many of the basins and craters formed during this period have long since disappeared, obliterated by subsequent impacts. Traces of one of them, the giant Procellarum Basin, can still be seen as Oceanus Procellarum (Ocean of Storms).

Nectarian Period   This period began when a giant asteroid carved out the Nectaris Basin, which later became Mare Nectaris (Sea of Nectar). Samples from Apollo 16 indicate that the asteroid struck approximately 3.92 billion years ago. Several other large basins were blasted out during this period, including Serenitatis, Crisium, and Humorum. These basins were, in effect, giant craters. The impacts that created them melted tremendous amounts of lunar material, and sent shock waves rippling through the molten matter. As this material rapidly cooled, the shock waves formed concentric mountain rings. Similar events were also taking place on the Moon's far side. Many large near side craters, for example, Clavius and Alphonsus, were blasted out in the Nectarian Period, which spanned some 70 million years.

Early Imbrian Epoch   The Early Imbrian Epoch was marked by two main events. The first was the cataclysmic impact that created the 750 mile (1,200 km) diameter Imbrium Basin 3.85 billion years ago. The second was a similar collision that struck the Moon's western limb (the left edge as seen from Earth), forming the Orientale Basin. These distinctive lunar features provided important evidence regarding the role of asteroids in shaping the lunar surface. They clearly showed how the energy released by giant impacts melted the Moon's surface, sending molten matter and concentric shock waves across its face.

The outermost mountain ring of the Imbrium Basin is marked by the towering Apennines (visited by Apollo 15), as well as the Alps, Caucasus, and Carpathian mountains. The large grooves that cut through the Apennines were carved by flying debris from the Imbrium impact. Some of the ejecta reached the Apollo 16 site in the lunar highlands 1,000 miles (1,600 km) from the impact point.

The Orientale Basin offers an even better example of the effects of large asteroid impacts on the Moon's surface. Our best views of this classic basin were provided by Lunar Orbiter 4 (the Apollo astronauts flew over the region in darkness). As the photograph on page 150 illustrates, Orientale resembles a giant bull's eye. Its multi-ringed mountain chains and furrowed terrain clearly reveal the effects of shock waves and flying debris.

Late Imbrian Epoch   The Early Imbrian Epoch lasted about 50 million years, after which the huge basin-making impacts came to an end. Large meteorites still punched craters into the Moon's face, but the asteroids that blasted out the lunar basins had apparently been swept clean from our region of the Solar System. It should be noted that Earth was also subject to asteroid strikes during that time, but the effects of surface erosion and plate tectonics have erased most traces of this violent period on our own planet.

However, conditions on the Moon were very different. Its thinner crust was cracked and weakened by the basin-making impacts, and lava from the mantle—the layer between the Moon's crust and its core—started flowing in successive sheets through countless cracks and fissures. This marked the beginning of the Late Imbrian Epoch, lasting some 600 million years from 3.8 to 3.2 billion years ago. Dark lava filled the circular basins, forming today's familiar lunar maria. On the near side, where the crust was the thinnest, the lava flows were the thickest. The far side's paucity of maria is due to its thicker crust which better withstood the basin impacts, preventing lava from reaching the surface. As heavy lava covered the near side maria, its weight further weakened the crust. As a result, the seas sagged at their centers, and cracked around their edges. That is why most of the lunar faults, ridges, and rilles are located in the maria. Some of the lava flowed though channels, such as Hadley Rille (visited by the Apollo 15 astronauts). Some of it erupted from geyser-like fire fountains, which shot hot magma high into the lunar sky. Cooling rapidly, much of this spray formed tiny glass spheres which fell back to the lunar surface, piling up around volcanic vents and across the surrounding terrain. That is what happened at Shorty Crater, where the Apollo 17 astronauts discovered the famous orange soil which was composed of orange and black beads of glass. Lava flowed over the Moon's face at different places in different times. Samples from the Sea of Tranquillity, for example, are 3.7 billion years old. The fragments brought home by Luna 24 from the Sea of Crises are 3.3 billion years old. Rocks from the Ocean of Storms are 3.2 billion years old. While magma continued to spread across the lunar surface after this time, the Late Imbrian Epoch marked the most active phase of volcanism on the Moon.

Eratosthenian Period   The frequency of large meteorite strikes, along with the level of volcanism, both rapidly declined during the Eratosthenian Period, which spanned 2.1 billion years from 3.2 to 1.1 billion years ago. Since this period, the Moon has changed very little. The ages of large craters from the period, for example, Eratosthenes, Langrenus, and Theophilus, can be deduced by their eroded appearance.

The Moon has been geologically dead for the past 3 billion years. By contrast, life still hadn't appeared on Earth three billion years ago!

Copernican Period   A steady rain of tiny meteorites and particles from the Sun has bombarded the Moon for billions of years, churning up the ancient crust, and creating the lunar regolith. However, few large craters have been blasted out during the last 1.1 billion years. Copernicus, one of the Moon's most prominent rayed craters, was formed about 800 million years ago (as determined by Apollo 12 samples). Tycho, a spectacular rayed crater, was formed about 100 million years ago. Some of the smaller craters visited by the Apollo astronauts are as "young" as two million years old.

The Apollo Lunar Surface Experiments Package (ALSEP) seismometers recorded several recent meteorite impacts on the Moon, and Jack Schmitt saw "a little pinprick of a bright flash" during Apollo 17, which must have been a meteor strike. But large objects hitting the Moon are now rare. It is estimated that only between one and three craters, six miles (10 km) or more in diameter, are formed every 10 million years.

Where did the Moon come from? Scientists now believe that a Mars-sized object hit the young Earth, giving birth to the Moon. The object struck our planet with a glancing blow, tearing the smaller body apart, and sending fragments into orbit around Earth. The debris eventually came back together through accretion to form the Moon, which then evolved as described in the above geologic history.

Commonly Used Acronyms

ALS.......................... Apollo Landing Site
ALSEP .... Apollo Lunar Surface Experiments Package
BPC .... Boost Protective Cover
CDR .... Commander
CM .... Command Module
CMP .... Command Module Pilot
CSM .... Command Service Module
EDT .... Eastern Daylight Time
EMU .... Extravehicular Mobility Unit
EST .... Eastern Standard Time
EVA .... Extravehicular Activity
IGY .... International Geophysical Year
ICBM .... Intercontinental Ballistic Missile
JPL .... Jet Propulsion Laboratory
KSC .... Kennedy Space Center
LC .... Launch Complex
LEM .... Lunar Excursion Module
LES .... Launch Escape System
LM........................... Lunar Module
LMP .... Lunar Module Pilot
LOI.......................... Lunar Orbit Insertion
LOR .... Lunar Orbit Rendezvous
LRL .... Lunar Receiving Laboratory
LRRR .... Laser Ranging Retro-Reflector
LRV .... Lunar Roving Vehicle
MCC .... Mission Control Center
MET .... Modularized Equipment Transporter
MSC .... Manned Spacecraft Center
MSFC .... Marshall Space Flight Center
NASA .... National Aeronautics and Space Administration
PLSS .... Portable Life Support System
RCS .... Reaction Control System
RTG .... Radioisotope Thermoelectric Generator
SIM .... Scientific Instrument Module
SLA.......................... Spacecraft-Lunar Module Adapter
SM .... Service Module
SPS .... Service Propulsion System
TEI .... Transearth Injection
TLI .... Translunar Injection
USGS .... United States Geological Survey
UV .... Ultraviolet
VAB .... Vehicle Assembly Building

Glossary

Accretion

The growth of a body in space by the coming together of many smaller bodies.

Alpha particle

A positive particle consisting of two protons and two neutrons. It is the nucleus of a helium atom.

Anorthosite

A rock composed primarily of the mineral plagioclase, believed to be the main constituent of the Moon's original crust.

Ascent Stage

The upper stage of the Apollo Lunar Module which carried the astronauts from the Moon's surface back into lunar orbit.

Asteroid

A rocky or metallic interplanetary body usually larger than 100 yards (91 m) across. The largest known asteroid is roughly 600 miles (1,000 km) in diameter.

Astrogeology

The science which studies the composition and geologic history of the Moon, planets, and similar bodies in space.

Basin

A huge depression in the Moon's crust formed by the impact of a large asteroid or comet.

Basalt

A fine-grained, dark gray, igneous rock formed by the solidification of magma (lava). The volcanic rocks found in Hawaii are basalts, as were the majority of rocks collected from the lunar maria.

Breccia

A coarse-grained rock composed of angular fragments of pre-existing rocks. On the Moon, breccias are fused together by the heat generated when meteorites strike the surface. Breccias were found all across the Moon, especially in the highland regions.

Caldera

A large volcanic crater.

Comet

A large interplanetary body composed of rock and ice. When heated by the Sun in the inner Solar System, comets release gas that form bright distinctive heads and tails. Comets were formed in the cold outer regions of the Solar System. Traveling in highly elliptical orbits, they spend most of their time far beyond Pluto.

Command Module The cone-shaped section of the Apollo spacecraft that housed the astronauts during most of their mission. It was the only part of the spacecraft that returned to Earth.

Crater A bowl-shaped depression, usually with an elevated rim. Most craters found on Earth were formed by volcanoes, while most craters found on the Moon were formed by meteorite impacts. When a large meteorite strikes the Moon and throws out large blocks of rock which fall back to the surface, *secondary* craters are formed.

Crust The hard outer layer of the Moon which encases an inner layer, the *mantle,* and possibly a small partly-molten core.

Descent Stage The lower stage of the Apollo Lunar Module which housed the Descent Engine for landing on the Moon's surface. It also served as a "launch pad" for the Ascent Stage.

Dome A small pimple-shaped feature on the lunar surface formed by rising magma. Hundreds of domes exist, mostly in the maria.

Earthshine The illumination of the lunar surface by sunlight reflected by Earth (as seen when the Moon is a thin crescent).

Ejecta Lunar material thrown out of a crater or basin by the force of a large impact.

Erosion The disintegration of surface features caused by natural forces, such as wind and water. Erosion on the Moon is mostly the result of meteorite impacts over millions of years.

Eon A period of geologic time equal to one billion years.

EVA Extravehicular activity (any activity by an astronaut outside of a spacecraft).

Far side That side of the Moon which always faces away from Earth. Never properly referred to as the *dark* side.

Fault A fracture or fracture zone in the Moon's crust which is responsible for the formation of many of the lunar rilles and scarps.

Fines

Small particles of pulverized rock and glass which comprise lunar soil samples.

Fire fountain

A geyser-like eruption of gas-charged lava from a volcanic vent that creates a fountain of fiery molten rock.

Gardening

The shattering, overturning, and changing of the lunar surface as a result of meteorite impacts and other processes, resulting in the formation of the lunar regolith (topsoil).

Highlands

The lighter portions of the lunar surface, as seen from Earth. Also referred to as *terra*. When the smooth dark maria were originally identified as "seas," the rough brighter parts of the Moon were referred to as "highlands." These regions, which cover most of the Moon's surface, are higher than the maria and represent the original lunar crust.

Isotopes

Elements having the same number of protons in their nuclei, but differing in the number of neutrons. Isotopes of a given element have the same general chemical properties, but differ slightly in mass and in some physical properties. An isotope has the same atomic number as its element (i.e. deuterium is an isotope of hydrogen, both having the atomic number of 1).

Lava

1. Flowing magma (molten rock) which reaches the surface of a planet through a fissure or volcano. *2. Magma,* after it has cooled and solidified.

Magma

Underground molten rock that may or may not contain suspended solids, such as rock fragments and crystals, or bubbles of gas.

Mare

plural *maria.* Any of the Moon's smooth dark areas, once thought to be lunar seas. Maria are relatively smooth plains, mostly on the Moon's near side, that were formed when successive lava flows filled or partly filled the lunar basins.

Meteorite

A small rocky or metallic interplanetary body that has struck Earth or the Moon. Meteorites range in size from microscopic (micrometeorites) to many tons. Their composition ranges from silicate rocks to metallic iron-nickel. While traveling through space, meteorites are called *meteoroids.* While falling through Earth's atmosphere, meteoroids are called *meteors,* and more commonly, *shooting stars.*

Mineral   A chemical element or compound, usually in crystalline form. Minerals are the substances that constitute rock.

Moon   *1.* Earth's natural satellite. *2.* Any natural satellite revolving around a planet (e.g., Europa is a moon of Jupiter).

Near side   That side of the Moon which always faces Earth. Because of the Moon's apparent rocking motion (an effect called *libration*), we can actually see some 59 percent of the total lunar surface area.

Ray   A bright streak of ejected material that extends radially from a crater on the Moon or other planet. Individual rays may be hundreds of miles (or kilometers) long.

Regolith   The powdery upper layer of fragmented and pulverized material which constitutes the Moon's topsoil. The regolith was produced by meteorite bombardment over billions of years; its depth ranges from 3 to 50 feet (1 to 15 m), depending on local conditions.

Rille   Also called *rima.* A long narrow lunar valley. There are two types of rilles: Linear rilles are thought to be grabens (narrow blocks of crust that have dropped between two parallel faults), while meandering sinuous rilles are most likely ancient lava channels or collapsed lava tubes that carried molten rock away from volcanic vents.

Rim   An elevated region or lip around a lunar crater.

Sample   A representative rock or bag of soil, collected from a specific location on the Moon and returned to Earth for analysis. The chemical composition and measured age of individual lunar samples have helped scientists determine the Moon's geologic history.

S-band   A range of radio frequencies used in communication between Earth and the Moon.

Service Module   The cylindrical part of the Apollo spacecraft which provided oxygen, water, and electricity to the Command Module (CM). Mated to the CM, the Service Module also housed the Service Propulsion System (SPS), whose engine placed the spacecraft into lunar orbit, and also sent the three astronauts back to Earth.

Solar wind   The continuous radial flow of particles emitted from the Sun. The solar wind extends beyond the outer planets. Near Earth, solar wind speed can reach 600 miles (1,000 km) per second. The solar helium-3 in the lunar soil was carried to the Moon by the solar wind over billions of years.

Terminator   The line separating the illuminated and darkened areas of a body that does not shine by its own light (i.e., the line that separates night from day).

Thrust   The force generated by a rocket engine. Thrust is measured in pounds or kilograms.

Volcanic vent   An opening or channel in the Moon's crust through which magma is transported, and out of which lava erupts at the surface.

Volcanism   Volcanic activity. The term usually includes all natural processes resulting in the formation of volcanic rocks, lava flows, craters, etc.

Wrinkle ridge   A sinuous elevation, up to 20 miles (32 km) wide and 100 yards (91 m) high, which can extend for hundreds of miles across the lunar maria. Wrinkle ridges are associated with the lava flows that filled the lunar basins, forming the maria.

Bibliography

Books, Booklets, Press Kits, and Articles

Benson, Charles D., and William Barnaby Faherty. *Moonport: A History of Apollo Launch Facilities and Operations.* Washington: NASA SP-4204, 1978.

Bilstein, Roger E. *Stages to Saturn: A Technological History of the Apollo/Saturn Launch Vehicles.* Washington: NASA SP-4206, 1980.

Borisenko, Ivan, and Alexander Romanov. *Where All Roads Into Space Begin.* Moscow: Progress Publishers, 1982.

Brooks, Courtney G., James M. Grimwood, and Loyd S. Swenson, Jr. *Chariots for Apollo: A History of Manned Lunar Spacecraft.* Washington: NASA SP-4205, 1979.

Carr, Michael H., ed. *The Geology of the Terrestrial Planets.* Washington: NASA SP-469, 1984.

Cherrington, Ernest H., Jr. *Exploring the Moon Through Binoculars and Small Telescopes.* New York: Dover Publications, 1984.

Collins, Michael. *Carrying the Fire: An Astronaut's Journey.* New York: Farrar, Straus and Giroux, 1974.

. *Mission to Mars: An Astronaut's Vision of Our Future in Space.* New York: Grove Weidenfeld, 1990.

Compton, William David. *Where No Man Has Gone Before: A History of Apollo Lunar Exploration Missions.* Washington: NASA SP-4214, 1989.

Cortright, Edgar M., ed. *Exploring Space With a Camera.* Washington: NASA SP-168, 1968.

. *Apollo Expeditions to the Moon.* Washington: NASA SP-350, 1975.

Crosscurrents, *The First Man in Space: The Record of Yuri Gagarin's Historic First Venture into Cosmic Space.* New York: Crosscurrents Press, 1961.

Drake, Stillman. *Discoveries and Opinions of Galileo.* Garden City, NY: Doubleday & Company, 1957.

French, Bevan M. *The Moon Book.* New York: Penguin Books, 1977.

Glushko, Valentin P., *Development of Rocketry and Space Technology in the USSR.* Moscow: Novosti Press Agency Publishing House, 1973.

Hacker, Barton C., and James M. Grimwood. *On the Shoulders of Titans: A History of Project Gemini.* Washington: NASA SP-4203, 1977.

Hartmann, William K. *Astronomy: The Cosmic Journey.* Belmont, CA: Wadsworth Publishing Company, 1991.

Heiken, Grant H., David T. Vaniman, and Bevan M. French. *Lunar Sourcebook: A User's Guide to the Moon.* New York: Cambridge University Press, 1991.

Kane, Phillip S., *Moon Observer's Planner: 1994.* Fillmore, CA: Skywatch Publishing, 1993.

Kosofsky, Leon J., and Farouk El-Baz. *The Moon as Viewed by Lunar Orbiter.* Washington: NASA SP-200, 1970.

Lehman, Milton. *Robert H. Goddard: Pioneer of Space Research.* New York: Da Capo Press, 1988.

Masursky, Harold, G.W. Colton, and Farouk El-Baz, eds. *Apollo Over the Moon: A View from Orbit.* Washington: NASA SP-362, 1978.

Mishin, Vasili, and Boris Raushenbakh. *The Scientific Legacy of Sergei Korolev: Selected Works and Documents.* Moscow: USSR Academy of Sciences, 1980.

Murray, Charles, and Cathrine Bly Cox. *Apollo: The Race to the Moon.* New York: Simon and Shuster, 1989.

NASA. *Apollo 8 Mission Report.* Washington: NASA MR-2, 1969.

. *Apollo 10 Mission Report.* Washington: NASA MR-4, 1969.

. *Apollo 12 Mission Report.* Washington: NASA MR-8, 1970.

. *Apollo 13 Mission Report.* Washington: NASA MR-7, 1970.

. *Apollo 14 Mission Report.* Washington: NASA MR-9, 1971.

. *Apollo 15 Mission Report.* Washington: NASA MR-10, 1971.

. *Apollo 17 Mission Report.* Washington: NASA MR-12, 1972.

. *Apollo 8 Press Kit.* Washington: NASA, 1968.

. *Apollo 9 Press Kit.* Washington: NASA, 1969.

. *Apollo 10 Press Kit.* Washington: NASA, 1969.

. *Apollo 12 Press Kit.* Washington: NASA, 1969.

. *Apollo 15 Press Kit.* Washington: NASA, 1971.

. *Apollo 8: Man Around the Moon.* Washington: NASA EP-66, 1969.

. *Apollo 9: Code-Name Spider.* Washington: NASA EP-68, 1969.

. *Apollo 11 Log.* Washington: NASA EP-72, 1969.

. *Apollo 13: "Houston, We've Got a Problem".* Washington: NASA EP-76, 1970.

. *Apollo 14: Science at Fra Mauro.* Washington: NASA EP-91, 1971.

. *Apollo 15: At Hadley Base.* Washington: NASA EP-94, 1971.

. *Apollo 16: At Descartes.* Washington: NASA EP-97, 1972.

. *Apollo 17: At Taurus-Littrow.* Washington: NASA EP-102, 1973.

. *Lunar Orbit Rendezvous: News Conference on Apollo Plans at NASA Headquarters on July 11, 1962.* Washington: NASA, 1962.

. *Lunar Orbiter Press Kit.* Washington: NASA, 1966.

. *NASA Facts: Lunar Orbiter.* Washington: NASA NF-32, 1967.

. *NASA Facts: Surveyor.* Washington: NASA NF-35, 1967.

. *NASA Facts: Manned Space Flight—The First Decade.* Washington: NASA NF-48, 1973.

. *Planetary Exploration Through Year 2000: Scientific Rationale.* Washington: NASA, 1988.

. *Report of the Apollo 204 Review Board.* Washington: NASA, 1967.

. *Status and Future of Lunar Geoscience.* Washington: NASA, 1986.

. *Surveyor A Press Kit.* Washington: NASA, 1966.

. *Surveyor G Press Kit.* Washington: NASA, 1968.

. *Surveyor Program Results.* Washington: NASA SP-184, 1969.

Nicks, Oran W. *Far Travelers: The Exploring Machines.* Washington: NASA SP-480, 1985.

Pannekoek, Anton. *A History of Astronomy.* New York: Dover Publications, 1989.

Pellegrino, Charles R., and Joshua Stoff. *Chariots for Apollo: The Making of the Lunar Module.* New York: Atheneum Publishers, 1985.

Petrovich, G.V., ed. *The Soviet Encyclopedia of Space Flight.* Moscow: Mir Publishers, 1969.

Riabchikov, Evgeny. *Russians in Space.* Garden City, NY: Doubleday & Company, 1971.

Rukl, Antonin. *Atlas of the Moon.* London: Hamlyn, 1991.

Schmitt, Harrison H. "Exploring Taurus-Littrow." *National Geographic.* Washington: pp. 290–306, September 1973.

. "Evolution of the Moon: The 1974 Model." *Space Science Reviews 18.* Dordrecht, Holland: D. Reidel Publishing Co., pp. 259–279, 1975.

Scott, David R. "What Is It Like to Walk on the Moon?" *National Geographic.* Washington: pp. 326–329, September 1973.

Shoemaker, Eugene M. "The Moon Close Up." *National Geographic.* Washington: pp. 690–707, November 1964.

Simmons, Gene. *On the Moon With Apollo 15: A Guidebook to Hadley Rille and the Apennine Mountains.* Washington: NASA, 1971.

. *On the Moon With Apollo 16: A Guidebook to the Descartes Region.* Washington: NASA, 1972.

. *On the Moon With Apollo 17: A Guidebook to Taurus-Littrow.* Washington: NASA, 1972.

Stuhlinger, Ernst, and Frederick I. Ordway III. *Wernher Von Braun: Crusader for Space.* Malabar, FL: Krieger Publishing Company, 1994.

Swann, Gordon A., et al. *Geology of the Apollo 14 Landing Site in the Fra Mauro Highlands.* Washington: U.S. Geological Survey Professional Paper 880, 1977.

Swenson, Loyd S., Jr., James M. Grimwood, and Charles C. Alexander. *This New Ocean: A History of the Project Mercury.* Washington: NASA SP-4201, 1966.

Ulrich, George E., Carroll Ann Hodges, and William R. Muehlberger. *Geology of the Apollo 16 Area, Central Lunar Highlands.* Washington: U.S. Geological Survey Professional Paper 1048, 1981.

Verne, Jules. *From the Earth to the Moon.* New York: Airmont Books, 1967.

. *Round the Moon.* New York: Airmont Books, 1969.

Von Braun, Wernher. *The Mars Project.* Urbana, IL: University of Illinois Press, 1991.

Von Braun, Wernher, and Frederick I. Ordway III. *History of Rocketry & Space Travel: Revised Edition.* New York: Thomas Y. Crowell Company, 1969.

Weaver, Kenneth F. "The Moon." *National Geographic*. Washington: pp. 207-228, February 1969.

. "What the Moon Rocks Tell Us." *National Geographic*. Washington: pp. 789-791, December 1969.

. "Have We Solved the Mysteries of the Moon?" *National Geographic*. Washington: pp. 309-325, September 1973.

Wilhelms, Don E. *The Geologic History of the Moon*. Washington: U.S. Geological Survey Professional Paper 1348, 1987.

. *To a Rocky Moon: A Geologist's History of Lunar Exploration*. Tucson, AZ: The University of Arizona Press, 1993.

Wilson, Andrew. *Solar System Log*. London: Jane's Publishing Company Limited, 1987.

Wolfe, Edward W., et al. *The Geologic Exploration of the Taurus-Littrow Valley: Apollo 17 Landing Site*. Washington: U.S. Geological Survey Professional Paper 1080, 1981.

NASA Films

NASA, *Apollo 8: Go For TLI*. Washington: NASA, 1969.

. *The Flight of Apollo 11: "The Eagle Has Landed."* Washington: NASA, 1969.

. *Apollo 12: Pinpoint for Science*. Washington: NASA, 1969.

. *Apollo 13: "Houston, We've Got a Problem."* Washington: NASA, 1970.

. *Apollo 14: Mission to Fra Mauro*. Washington: NASA, 1971.

. *Apollo 15: In the Mountains of the Moon*. Washington: NASA, 1971.

. *Apollo 16: "Nothing So Hidden…"* Washington: NASA, 1972.

. *Apollo 17: On the Shoulders of Giants*. Washington: NASA, 1973.

Maps

The Earth's Moon. Washington: National Geographic Society, 1976.

USGS, *Geologic Map of Apollo Landing Site 2 (Apollo 11)*. Washington: U.S. Geological Survey I-619, 1970.

. *Geologic Map of the Lansberg P Region of the Moon*. Washington: U.S. Geological Survey I-627, 1971.

. *Geologic Maps of the Fra Mauro Region of the Moon*. Washington: U.S. Geological Survey I-708, 1970.

. *Geologic Maps of of the Apennine-Hadley Region of the Moon*. Washington: U.S. Geological Survey I-723, 1971.

. *Geologic Maps of the Descartes Region of the Moon*. Washington: U.S. Geological Survey I-748, 1972.

. *Geologic Maps of the Taurus-Littrow Region of the Moon*. Washington: U.S. Geological Survey I-800, 1972.

Acknowledgments

This book would not have been possible without the help and support of the many people who provided research materials, photographs, interviews, and encouragement. I am grateful to all of them, especially to:

My parents, *Frank* and *Lorraine Mellberg,* who sparked my interest in the Moon in 1963 by giving me a telescope for my eleventh birthday. My father also kept me informed about his work on Project Surveyor, while my mother fostered my early interest in reading, including science books, and Jules Verne's prophetic fictional tales.

William J. O'Donnell, former Public Affairs Officer for Manned Space Flight at NASA Headquarters in Washington, with whom I started corresponding 30 years ago during the space program's heyday. Bill patiently answered every one of my letters and sent me press kits, fact sheets, and photographs, many of which were used in preparing this book. Now retired, Bill still answers my letters, and he reviewed my manuscript, offering some excellent suggestions for improvements.

Dr. Harrison H. "Jack" Schmitt, Apollo 17 astronaut, astrogeologist, former United States Senator, and, in my opinion, one of the most articulate Moonwalkers. Dr. Schmitt generously granted me interviews in 1994 and 1996. I am honored that this kindly and visionary man was also willing to contribute the Foreword to *Moon Missions.* His ideas regarding the future of lunar exploration and industrialization are thought-provoking and bring a unique perspective to this book. My respect for him is enormous.

Steven Jay, friend, partner, and radio personality, who set up and hosted the interviews with Jack Schmitt. Steve's enthusiasm and encouragement have helped me tremendously.

James C. Floyd, one of the world's foremost aerospace engineers, who told me about Canada's many contributions to America's space program. Over the years, Jim has been more than a mentor—he's been an inspiration.

Vladimir Lytkin, Scientific Director of the Tsiolkovsky Cosmonautics Museum in Kaluga (Russia), and *Boris Belitzky,* Science Editor at Radio Moscow, who provided details and photographs of the Soviet Union's space program.

C. Frederick Matthews, one of NASA's original flight controllers, who shared his memories of Project Mercury.

Lawrence C. Freudinger, one of NASA's current engineers, whose farsighted ideas are greatly appreciated.

Dick and *Paula Eastman,* who offered their help in so many ways.

Dennis Marshak, Mark Dressler, and *all of their colleagues* at Photochrome Prints in Park Ridge, Illinois, who helped to restore and reproduce some rare photographic materials.

Jerry Berg, Becky Fryday, Lynda Matys, Jurrie van der Woude, Nena Wilson, and the public affairs offices at NASA Headquarters and outlying NASA Centers.

Last, but certainly not least, I want to thank my publisher, *Plymouth Press,* and particularly my editor, *Jan Jones,* for their faith in this project, and their countless hours of hard work in pulling it all together. Their suggestions and support were invaluable.

Index

About the Author

Bill Mellberg is a well-known aerospace writer and historian. His first book, *Famous Airliners*, traced the development of civil air transportation, while his articles in aerospace publications have reached audiences around the globe. He is a former public relations representative for Fokker Aircraft. Bill is also a nationally-recognized political satirist and speaker who has starred in several television specials for PBS. A native of Chicago, Bill was graduated from the University of Illinois at Urbana-Champaign in 1975. He is a member of The Planetary Society and of the Association of Lunar and Planetary Observers. You'll find him during quiet evenings — when he can find the time — viewing the Moon and other celestial objects through his backyard telescope.

If you enjoyed Moon Missions, then you'll also enjoy Bill Mellberg's other book, *Famous Airliners*...

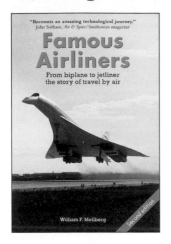

Famous Airliners

The evolution of the modern airliner is chronicled in *Famous Airliners*. Its story begins with the Boeing Model 80, whose 12 passengers flew in the "Pioneer Pullman of the Air" at a little over 100 mph and were the first to enjoy the attention of specially trained nurses called "stewardesses." It ends with the Concorde, which travels at more than twice the speed of sound and over the Atlantic in about three hours. Includes over 70 color and black and white photos of vintage and modern aircraft.

7" by 10", 196 pages

$19.95 ISBN 1-882663-13-6

Guide to Airport Airplanes

Employing a simple-to-use, systematic approach, *Guide to Airport Airplanes* facilitates rapid identification of airliners. Features the 66 most commonly observed airliners with a capacity of 19 or more passengers, all pictured in full-color photographs while in flight or at interesting airport locales. Airliner capabilities such as cruising speed, range, and passenger capacity, as well as country of origin and date of first flight are included.

7" by 5", 168 pages, 70 color photos

$14.95 ISBN 1-882663-10-1

"A handy guide that can be taken along on trips for quick identification." **Toronto Star**

We commisioned celebrated aviation artist Richard King to create six new airliner paintings especially for these note card sets. Modern airliners make up one set, while beloved vintage airliners are depicted in the other. Note cards are of heavy stock, measure a full 5 by 6⅝ inches folded, and are packed in attractive, sturdy boxes *(box covers shown below)*.

Jetliners

Vintage Airliners

Modern Jetliners set *includes*
9 note cards and envelopes
3 each of 3 paintings: Boeing 747
(*shown above*), McDonnell-Douglas
DC-9, and Airbus A300
$9.95 each boxed set

Vintage Airliner set *includes*
9 note cards and envelopes
3 each of 3 paintings: Douglas
DC-3 (*shown above*), Martin
China Clipper, and Ford Trimotor
$9.95 each boxed set

To order, or for a free catalog of aviation books & gifts, call (800) 477-2398.
MasterCard, VISA, and American Express accepted.